FORM FORESTRY

JOHN ARDEN FERGUSON

GREENWORLD
PUBLISHERS

First Indian Reprint 2001

ISBN 81-87421-21-5

Published by
Greenworld Publishers
8/217, Indira Nagar
Lucknow - 226 016 (UP)
Tel. : 701433

Exclusively Distributed by
International Book Distributing Company
Chaman Studio Building, 2nd Floor,
Charbagh, Lucknow - 226 004 (India)
Tel. : Off. : 450004, 450007, 459058 Res. : 711172, 342829
Fax : 0522-458629
E-Mail : ibdco@sancharnet.in

Printed at :
Army Printing Press
33, Nehru Road, Sadar Cantt
Lucknow - 226 002
Tel. : 0522- 481164, 480546
E-mail : pressneversleeps@rediffmail.com

To Her

WHO LED ME TO

BELIEVE IT POSSIBLE

TO WRITE A BOOK

THAT MIGHT BE

OF SERVICE

PREFACE

THIS book is an outgrowth of lectures delivered to Agricultural students on Farm Forestry throughout several years. It is written for study by students in Agricultural Colleges and in High Schools.

In the preparation of the book the author owes much to the many writers on subjects relating to the care and management of the farm woodlot and the utilization of the products of the woodlot. The Bulletins and Circulars issued by the United States Forest Service, by Agricultural Experiment Stations and by State Forestry Departments relating to these subjects, are full of valuable material and are the chief sources of information. The aim has been to bring together in available form ideas and principles already well known.

An appendix has been added containing several log tables and volume tables, a basal area table, and other information, nearly all taken from Bulletins of the United States Forest Service. A simple method for finding the area of a farm woodlot using the common compass has been described, and a suggestive list of practicum exercises such as might be included in a course on Farm Forestry. Photographs not otherwise acknowledged were obtained from the United States Forest Service.

The author wishes to acknowledge his indebtedness to the following, all of whom made valuable suggestions and corrections in the manuscript: R. C. Hawley, who read chapters 1-11; H. H. Chapman, who read Chapters 11-18, both of the Faculty of the Yale Forest School; A. K. Chittenden, Professor of Forestry at the Michigan Agricultural College, who read Chapters 1-18, and to my associates on the Faculty of the Department of Forestry, of The Pennsylvania State College; C. R. Anderson, who made many valuable suggestions; R. R. Chaffee, who read Chapters 11-20; W. G. Edwards, who helped in the preparation of the index, and G. R. Green, who read and corrected the entire manuscript.

J. A. FERGUSON.

CONTENTS

PAGE

CHAPTER I

The Origin and Uses of the Woodlot 1

CHAPTER II

The Tree 10

CHAPTER III

The Life Processes of Trees and the Natural Factors that Influence the Life of Trees 20

CHAPTER IV

The Woodlot 29

CHAPTER V

The Life History of Trees in the Woodlot 36

CHAPTER VI

Establishing a Woodlot by Artificial Means, either by Seeding or Planting 45

CHAPTER VII

Trees to Use in Starting Woodlots 56

CHAPTER VIII

Trees and their Value for Planting 73

CHAPTER IX

Natural Methods of Starting and Reproducing a Woodlot 97

CHAPTER X

Caring for the Growing Woodlot 104

PAGE

CHAPTER XI

Caring for the Woodlot Started from Sprouts and the Ordinary Farm Woodlot 114

CHAPTER XII

Protection of the Woodlot 126

CHAPTER XIII

Woodlot Management 138

CHAPTER XIV

The Products of the Woodlot 145

CHAPTER XV

Harvesting the Products of the Woodlot 154

CHAPTER XVI

Marketing the Products of the Woodlot 163

CHAPTER XVII

The Contents of Logs and Trees 170

CHAPTER XVIII

Estimating the Trees in the Woodlot for Board Feet and for Cordwood 180

CHAPTER XIX

The Different Properties of Woods 189

CHAPTER XX

The Use of Perishable Woods for Fence Posts 198

Appendix 207

Index 235

FARM FORESTRY

CHAPTER I

THE ORIGIN AND USES OF THE WOODLOT

By the term Farm Woodlot is meant the tract of wooded land which is maintained primarily to furnish fuel, posts, lumber and other wood products for use on the farm. The term is also used to denote larger tracts of forest land owned by farmers which are cut from time to time to secure a revenue.

Farm Forestry treats of the growing of forest trees in the woodlot and with the utilization of the products of the woodlot. It has to do with the starting of the trees from seed or from sprouts, with the tending of the trees as they grow, so that they will put forth their best and fastest growth, and with the harvesting and utilizing of the trees when ripe. It aims to produce the greatest amount of valuable timber in the shortest time possible. It accomplishes its purpose by growing in the woodlot only fast growing kinds of trees, by growing trees that will yield only a useful and valuable product, by giving to each tree the right amount of space it needs for growth and by protecting the trees from their many enemies.

Origin of the Farm Woodlot.—Most farm woodlots have descended directly from the virgin forest which originally covered so large a portion of the country. This virgin forest has disappeared for the most part in the eastern part of the country. Only isolated patches of the old timber remain. These forests were cut to convert the trees into lumber or to clear the land for agriculture. Usually not all the land denuded of its trees to make farm land could be cultivated. On nearly every farm area there were places here and there where the soil was too thin or the land too rough or stony or too steep to raise a farm crop. Those parts of the farm often were not entirely denuded of timber or if cleared were allowed to grow up again to forest trees, and so formed the woodlots. In the older settled parts of the country farm woodlots have been cut over many times and allowed to grow up again. In New England some of the woodlots are said to be in the seventh generation from the virgin forest.

Some farm woodlots have originated also from the abandoning of land once tilled. In the early days land would sometimes be cleared and used for growing crops that later would be found not suited for that purpose. Such lands were abandoned and soon would grow up naturally to forest trees. Farm woodlots also originate by planting trees or sowing the seed of trees. In the central portions of the country, on the prairies, where trees do not naturally reproduce themselves in competition with the heavy grass, this is commonly the origin of the farm woodlot.

Relation of Farm Forestry to General Forestry.—Farm forestry is a branch of the general subject of forestry. It comprises the principles of forestry that apply to the management of the farm woodlot. Forestry in general has to do with the management of all forest land. Many are apt to think of forests as tracts of wooded land covering large areas, and to think of forestry as applicable only to such extensive areas. Yet a woodlot is forested land and the fundamental principles of forestry that apply to a large timber tract, such as a National or a State Forest, apply equally well to the farm woodlot. There is no owner of forest land who is better able to practice forestry than the farmer on his woodlot. He has the time to devote to the work in the winter and his efforts will cost him nothing. The principles of forestry as they apply to the woodlot are easily learned and are simple in application. One who understands the principles of growing and caring for agricultural crops can readily comprehend the principles of growing forest crops.

Agriculture and Forestry.—Forestry does not differ widely from agriculture in its general purpose. They both have to do with the growing of crops from the soil. The essential difference is in the kind of crop and in the time it takes to mature it. Forestry has often been defined as the branch of agriculture that has to do with the growing of crops of forest trees. Forestry and agriculture both deal with the establishing, tending and harvesting of crops. With agriculture these three operations are usually accomplished in the same season or year. With forestry many years must elapse between the time the crops of trees are established and when they are ripe for the ax. The same laws of nature however are relied on to grow both kinds of crops.

Like agriculture, forestry can do nothing more than to bring about the best conditions for the growth of the crop. It is impossible to make farm crops or trees grow faster than nature has determined they shall grow. All the farmer accomplishes by his careful tilling and preparation of the soil is to bring about the best conditions he can for the reception and germination of the seed and the development of the crop. Forestry aims to aid nature by giving to each tree the conditions necessary for its best development.

An intimate relation exists between forestry and agriculture in the use of the soil. If all the land on the farm were fertile there would be little need for farm forestry. Usually some portions of the land will be too poor to grow farm crops. Forestry has to do with the utilization of such poor land. It aims to grow a crop of forest trees where it is impossible to utilize the land for other and more valuable farm purposes. Forestry and agriculture thus go hand in hand in the use of all the land on the farm. Agriculture is interested in the use of the fertile soil for the growing of valuable farm crops. Forestry concerns itself with the use of the poor soil on which the only profitable crop is forest trees.

The Value of the Farm Woodlot to the Country.—The area of woodlot and idle land in the country owned by farmers is very large. It was estimated in 1915 by the United States Department of Agriculture to be not less than 200,000,000 acres. This is an area larger than all the National and State Forests. It is larger than the states of Minnesota, Michigan, Wisconsin and Maine. The proper use of this vast aggregate area is of considerable importance not only to the owners but also to the country. It is probable that this area may be somewhat reduced in future years. East of the Mississippi River two-thirds of the farm woodlots lie in the South, where much of the land at present timbered will eventually be cleared for farm purposes as the country becomes more thickly settled. In the older settled regions probably the greater part of the fertile land is already being utilized for growing farm crops. The woodlot land and the farm land are there pretty definitely settled. It is probable that some of the poorer land now being tilled may gradually be abandoned as it is found to be more valuable for growing trees than for agricultural crops. It will be many years before the land suited for agriculture and that suited for forestry will be definitely determined. In spite of changes that may occur the area in farm woodlots in the country will always be large and the production of timber on this area will be of great economic importance. It should be used so as to contribute to the national wealth and prosperity. The productive capacity of the 200,000,000 acres of farm woodlots is enormous. It is estimated that if only fairly well stocked with forest trees, that is about 3000 board feet to the acre, the annual growth under forest practice would more than equal the annual lumber cut of the entire country, which is about 40 billion board feet. In other words the present area of farm woodlots under proper care and forest management could supply the country indefinitely with lumber. The present stand of timber on these woodlots is very large. It is estimated to exceed 250 billion board feet of saw timber and 1½ billion cords of cordwood. It is because of the great value of the farm woodlot to the nation as well as to the individual farmer that the woodlot problem is of importance.

Value of the Woodlot to the Farm.—A well stocked thrifty woodlot enhances the value of the farm property. Instances are not rare where farms have been purchased and the sale of the products from the woodlot when cleared have more than covered the cost of the farm. The woodlot serves as a reserve fund on which the owner of the farm can call at any time for financial backing. The purchase of fuel wood, posts and other products means the throwing away of profits on these necessary articles. In addition to this is the effort and expense of hauling the material an increased distance. Grown in the farm woodlot this material would cost nothing except the annual taxes and the interest charges together with the actual cost of getting the products out of the woodlot. The profits would remain in the hands of the owner.

Woodlots on Fertile Soil.—Where woodlots are grown on fertile soil the point is soon reached where it may be more profitable to do away with the woodlot and purchase wood supplies than to maintain the woodlot. But if the woodlot is increasing at the rate of one and a half to two cords per acre per year, which will be possible on fertile soil with well stocked woodlots of thrifty and fast growing trees, it often will be more profitable to maintain the woodlot on fertile soil than to grow an ordinary farm crop. In the central western states where trees are not plentiful and fence posts must be brought long distances at considerable expense for freight, plantations of trees for growing fenceposts have often been found to yield a higher annual profit than the same kind of land devoted to raising farm crops. As a rule the woodlot occupies the poor soil on the farm, and on such soil the only profitable crop is forest trees. On such situations the only thing that would compete with the woodlot would be grazing. Even where more might be made out of the soil in other ways, the convenience of having wood material at hand whenever needed is worth something to the farmer. The fact that a well stocked woodlot is a valuable asset that is readily turned into money or on which money can be borrowed in case of necessity, will often outweigh other considerations in the confidence it gives the owner.

The Size and Location of the Woodlot.—The size and location of the woodlot will depend on several factors such as the home need of the farm, the character of the land, and the market for the material from the woodlot. Most farm woodlots are the result of the clearing of the land for farm purposes. As the land was needed for field crops the forest was gradually pushed back until the present farm woodlots have resulted. So there is rarely a definite relation between the size of the farm and the size of the woodlot. The woodlot is not the result of planning and calculation and is not the result of the study of the wood needs of the farm. Usually the woodlot occupies the poorest land on the farm and is large or small according to the amount of such land. It is held by many that one-eighth of the farm should be in woodlot, which should supply the farm with all wood products needed and also furnish some products for sale to help pay for taxes, interest and protection. While such an area of wooded land might be sufficient for some farms, for others it might be too large or too small. Because of the great variation in the rate of growth in trees on different soils and situations, the area of woodlot that will just supply the right amount of material would have to be determined for each farm.

There is no definite relation between the location of the farm woodlot and the farm buildings and fields. It is usually located wherever the soil is too poor for cultivation. When the woodlot is on good soil, it will usually be found located at some distance from the homestead. The area lying immediately around the home and the barn would be the first to be cleared and the woodlot would usually occupy the land farther away. Where woodlots are in the nature of windbreaks or shelter-belts there is a definite relation of position in reference to the buildings, pastures, fields or orchards that are to be protected. This position will vary in different portions of the country depending on the direction of the cold or drying winds, the shelter belts being placed to windward of the area to be protected.

THE USES OF THE FARM WOODLOT

The Production of Wood Material for Farm Use.—The woodlot should furnish each year all the wood needed on the farm. It should provide fuelwood, fenceposts, poles, stakes and lumber for repairing buildings.

The Production of a Revenue.—The woodlot should bring in an income each year. Besides furnishing wood material for use on the farm it should pay for taxes, interest on the investment, and other expenses of the woodlot, by growing some products for sale as sawlogs, ties, poles, firewood or posts. Idle land should be planted to forest trees. No land on the farm should be unprofitable.

Utilization of Teams and Help in Winter.—A good sized woodlot will furnish work for both teams and help in winter when it is often a problem to know how to employ them profitably. Farming and lumbering make a good combination, the one utilizing the teams and help when work is slack with the other. It is easier to secure and keep good help when there is work the year round.

Windbreaks and Shelterbelts for Protection.—Woodlots often take the form of windbreaks or shelterbelts for protecting the farmstead, the growing crops or the stock from the wind. On the great plains some protection from the wind is necessary. Trees are planted to windward of the area to be protected either in a single row or in two or more rows. Besides shelter they may furnish fuel, posts and other wood supplies to the farm. Woodlots are often used to furnish protection to stock from the hot sun in summer and the cold winds of winter.

Protection of the Water Supply and to Prevent Erosion to the Soil.—Trees by their dense crowns protect the soil from the hot sun's rays and from winds and are often the direct cause of the presence of springs. The removal of trees around a spring has often been known to dry up the spring. The roots of trees spreading in all directions serve to prevent the soil from being washed away by rain or floods in spring. Erosion can often be stopped by planting trees. Woodlots are often maintained on land adapted to agricultural crops because of the danger from erosion.

As a Sugar Bush.—In the North East where the hard maple forms a large part of the woods, woodlots are often maintained for the production of maple sugar. The tapping of the trees does not injure the wood for lumber or any other purpose. A considerable revenue results from a well stocked sugar bush.

As a Reserve Fund.—A well stocked woodlot is in the nature of a reserve fund on which the owner can draw at any time. It furnishes a basis for borrowing money when needed or it can readily be turned into cash by cutting the trees and converting them into lumber.

For Rejuvenating Wornout Land.—In some regions, especially in the South, wornout land is allowed to grow up to forest trees. After the trees have grown to pole size the soil will be found to be fertile again and can be tilled after the trees have been cut. A woodlot if properly handled benefits and enriches the soil by the accumulation of leaves and twigs which decay and add fertility.

As Game Preserves.—Woodlots are the breeding places of many birds that help to keep down destructive insects. The cause of the great increase in insects and the damage they do is said to be the killing of the birds.

A woodlot of oak and hickory descended from the virgin forest, as shown by the stumps of old trees.

A woodlot of white pine twenty-two years old started artificially by planting small trees.

To Temper the Climate.—A woodlot has an influence locally on the climate. The air in the woodlot is cooler in summer and warmer in winter than that outside. It is also warmer at night and cooler in the daytime. These differences in temperature have a tendency to set up local breezes that help to moderate the extremes of weather. More rain falls in a wooded region than in one not wooded.

Woodlots Have an Esthetic Value.—The esthetic value of woodlots is hard to measure. A wooded region is more pleasing to the eye than one without trees. There will be a greater demand for farms where woodlots are cared for and this will have an effect on the value of the farm. Well cultivated fields, thrifty woodlots and good roads make the country side attractive. A well kept woodlot adds greatly to the value of farm property because of its appearance setting off the growing crops. A farm should be made attractive and pleasant so that anyone would enjoy living on it.

CHAPTER II

THE TREE

In order to manage a woodlot intelligently and successfully it is necessary first to know something about the life and structure of trees, and also to understand something of the factors that influence their growth and development.

Trees are living organisms, that is, they carry on certain life processes such as breathing, transpiration, digestion and assimilation of food material. They are the highest type of vegetation found on the earth, that is, they have the most complex structure and life history of any plant growth. Trees differ from herbs in having woody structure and perennial life and from shrubs in developing single stems that elevate the crowns above other forms of vegetation.

THE PARTS OF A TREE AND THEIR USES

The Roots.—A tree is made up of three main parts—roots, stem and crown. The root system of a tree is composed of larger and smaller roots, each serving a definite purpose. They may be compared in form to the large and small branches in the crown. The large roots tend to hold the tree firmly in the ground and prevent its being blown over by the wind. They conduct water and salts absorbed from the soil by the smaller roots. These substances are absorbed by root hairs that grow near the ends of the roots. As the roots grow and force their way through the soil the root hairs die and are replaced by new ones formed nearer the tips of the growing roots. It is through these small root hairs, far out from the stem of a tree, often beyond the ends of the branches, that a tree takes in most of its food from the soil.

A woodlot of white pine seeded naturally on an abandoned field.

A pine woodlot used for wood and for shelter.

The form of the root system varies with different species of trees. Some trees, like most of the oaks, the hickory and walnut, develop a tap root or single main root that grows down deep into the soil. Such trees are naturally wind firm. They are practically never uprooted. They will be broken off above ground before they will be blown over. With many other trees like the spruce and birch, the roots tend to spread out laterally in the upper layers of the soil. Such trees not having a firm hold on the ground are not windfirm. A third form of root system is where several strong lateral roots extend down obliquely deep into the soil like the maple and beech. These are fairly windfirm trees. The degree of windfirmness varies in the same kind of tree according to the conditions under which the tree is grown. Trees in the open are rarely seen uprooted no matter what kind of a root system is naturally developed. A tree is able to strengthen its root system to withstand the wind. Where trees grow closely together in the forest they rely on each other for protection. Each tree helps to hold the others up by preventing the wind from blowing the tree over too far. Strong root systems are not developed. If such trees are thinned out suddenly, many of the trees being removed, so that those left are deprived of their former support, they will be apt to be blown over easily. If such trees are thinned out through a series of years, the support given by neighboring trees being gradually removed, the trees left will thicken and strengthen their root systems and become more and more wind firm, until finally they will be able to stand alone in the heaviest winds.

A few trees will adapt their form of root system to the conditions of the soil on which they grow. With a deep soil a deeply penetrating root system is produced, while in more shallow soil a more spreading root system is formed. This is true of but few trees. In general it may be said that unless the soil is deep enough for the development of the characteristic form of root system, the growth of the tree will be slow and its form more or less irregular or stunted. The root systems of trees can be studied on those upturned by the wind.

The Trunk, Stem or Bole.—The trunk of the tree supports and elevates the crown. It conveys water and salts absorbed by the roots to the crown and serves as a store house for reserve food materials. The food manufactured by the leaves in any year is not always completely used up in that year for the growth of the tree or the ripening of fruit. Usually more food is made than is needed. This excess of food is stored in the tree over winter for use in the spring. Seed years are brought about by the gradual accumulation of reserve food materials in the tree.

From an economic standpoint the stem is the most valuable portion of a woodlot tree. On its size and shape depend the amount and kinds of wood products such as lumber, cordwood and posts that can be obtained.

Each species of tree has its characteristic form of stem. Some trees produce naturally a long, cylindrical stem free of branches, while others develop a short, tapering stem more or less branched or crooked. Trees grown in the open usually branch low down forming a thick, short trunk. The same kinds of trees grown in the forest produce long, cylindrical trunks often free of branches for many feet. The bark covering the stem is also characteristic of each species of tree in markings, thickness and color. It serves as a protection against cold, heat, mechanical injury and the entrance of insects and disease. It is impervious to the passage of liquids and so prevents the sap leaving the tree freely through the branches or stem.

The Crown.—The crown is usually the most conspicuous part of the tree. It is composed of branches and leaves together with buds, flowers and fruit. Many life processes are carried on in the crown. By their almost endless divisions the branches enable the tree to present a large extent of leaf surface to the action of the sun and air. Unlike the roots the branches have a more or less regular arrangement.

Each species of tree has its characteristic form of crown and habit of branching. In the open the kind of tree can often be told from the distance by the form of crown, as elm by its vase shaped crown, maple by its oval crown, spruce by its conical crown. The form of crown produced by a tree in the woods differs from the form of crown produced by the same kind of tree grown in the open. This is due to crowding by neighboring trees, which restricts lateral development. The form of crown developed under forest conditions is as characteristic of the species as the form in the open. The form of the crown varies also with the age of the tree. Trees in early life have a more or less conically shaped crown due to rapid height growth. In old age the crown widens out and forms a flat or rounded top. There is an intimate relation between the size of the crown and the extent of the root system. If for any reason the one is not allowed its full development it will have a corresponding effect on the other portion. This does not mean that there is a definite connection between a particular root and a certain branch. It is not true that when a root is cut off a branch fed by that root dies. Neither is it true that the roots on one side of a tree feed only the corresponding side of the crown, and that by fertilizing the ground on one side of a tree the crown can be made to extend in that direction. All the roots of a tree help to obtain water and salts from the soil for the use of the entire tree.

The leaves may be said to be the lungs and stomach of the tree. In them are carried on respiration, transpiration, absorption of food from the air and the manufacture of food materials. The shape and size of leaves vary in different trees, each kind of tree having a different form. They vary from the broad, flat leaves of hardwoods to the long, needlelike leaves of conifers or evergreen trees. Even on the same tree there is considerable variation in the shape and size of leaves.

Trees that shed their leaves in the fall as do most hardwood or broadleaf trees are called deciduous trees. Those that hold their leaves over winter as do most coniferous trees like the pine, spruce and hemlock, are called evergreen trees. Some evergreen trees retain their leaves for two years, others for three or four years and others for longer periods. Evergreen trees shed one year's growth, the oldest, every year. There are many evergreen broadleaf trees like the live oak and holly and also several deciduous conifers like the larch and bald cypress. The arrangement of leaves on the twigs is such that the leaves will receive the greatest amount of sunlight. In some trees the leaves are alternate on the twigs, in others in pairs on opposite sides of the twigs, each successive pair alternating in its position on the twig with the pair above. In many species the leaves are arranged in whorls.

The winter period is the rest period in the life of trees. Growth practically ceases in the early fall. Trees during the winter period are in a dormant condition, but many of the life processes including respiration and transpiration take place the same as in summer, only in a much less active manner.

The Buds.—Buds contain the embryo of the next year's twigs, leaves or flowers. They have regular arrangement on the stem like the leaves. Buds are formed in the latter part of the season's growth after the leaves have matured and the new twigs have about finished their growth. They are covered with layers of scales for protecting the growing parts from the cold and drying winds of winter. Buds are either terminal or lateral. The terminal buds, when present, are for the purpose of prolonging the branch the following spring. Lateral buds occur in the axils of the leaf stalks. Some of them produce leaves in the spring and others produce flowers. Flower buds can often be distinguished by their larger size and form. Adventitious buds are formed under certain conditions from the young, living wood of trees. When a tree is cut the sprouts

European larch grown in the open with room for the development of the branches. Short trunks and wide spreading crowns result.

European larch plantation. The crowded stand has prevented the development of the side branches. Rapid height growth results with long trunks and narrow crowns.

that grow from the stump arise from adventitious buds. They also give rise to the branches that start on the trunk of a tree when it has been opened up to the light by the removal of other trees or by pruning. Dormant buds are buds that do not develop immediately after being formed. They lie just beneath the bark and may wait many years before starting to grow.

The Flowers of Trees.—Trees have flowers the same as other plants that produce seed. Usually they are small and inconspicuous and so are rarely observed. Some forest trees are grown as ornamental trees because of their beautiful blossoms as the locust, the catalpa, the tulip, the black cherry and horse chestnut. Flowers on trees give rise to the fruit or seed by the process of pollinization the same as with flowering plants. In some trees as with the black cherry the flowers are perfect, that is, both pistils and stamens occur in the same flower. Some trees like the pines, chestnut, and oaks have the pistillate and staminate portions in separate flowers but both kinds of flowers on the same tree. Still others like the cottonwood, tree of heaven, ashes and willows have these flowers on different trees.

Fruit and Seed.—The fruit and seed are the result of the fertilization and maturing of the pistillate flowers. Fruit and seed vary greatly in form and size on different kinds of trees. Some seed is very small and light and is easily blown about by the wind such as poplar and willow seed, while other seeds are large and heavy like the walnut and hickory. Some seed has a fleshy covering like cherry, hackberry and cedar which are eaten and scattered by birds.

CLASSIFICATION OF TREES ACCORDING TO SIZE

On a basis of size trees are often divided into:

a. Seedlings—young trees up to 4 feet in height.
b. Small saplings—trees from 4 feet to 10 feet in height.

c. Large saplings—trees from 10 feet in height until they have reached a diameter of 4 inches 4½ feet from the ground.
d. Small poles—trees from 4 to 8 inches in diameter.
e. Large poles—trees from 8 to 12 inches in diameter.
f. Standards—trees from 1 to 2 feet in diameter.
g. Veterans—trees over 2 feet in diameter.

The Structure of Wood.—The roots, branches and stems of trees are composed of wood. The structure of wood is not simple like iron or gold, but like all organic material it is made up of minute cells. These are somewhat similar in appearance in cross section to the cells in honeycomb though many times smaller. Usually they cannot be distinguished without the aid of the magnifying glass. The walls of these cells are made up of cellulose which is composed largely of carbon. In the living state these cells are filled with a living substance called protoplasm. The cells are not all alike in wood even in the same kind of tree, but differ in form and size and in the thickness of cell walls. They also differ in their uses. Some cells conduct food material to or from the crown, some store away reserve food material until it is needed by the tree, some serve simply to give strength to the woody tissue, while others have the power of division into two or more cells and so produce growth in trees. Water forms a large part of the wood of a growing tree. Over half the weight will be composed of water chemically combined, and in addition to this the tree will contain from forty to sixty-five per cent of its dry weight of free water. This free water largely evaporates when a tree is cut, so that dry wood is much lighter in weight than green wood. Well seasoned wood contains from ten to twelve per cent of free water.

The Parts of the Stem of a Tree.—If the top of the stump of a recently cut oak tree is examined, it will be found to be made up of several parts. Around the outside will be observed the bark. This is in two layers, an outer protective corky layer and an inner, light colored, softer layer sometimes called the bast. Inside the bark is a narrow, light colored portion of wood called the sapwood, and inside this extending to the center of the tree, a darker portion called the heartwood. In the center of the tree in some species will be found the pith. Running from the bark toward the center of the tree narrow lines of woody tissue will be seen, some extending even to the pith. These are the pith rays or medullary rays. In some trees like the oak, beech and sycamore the pith rays are very conspicuous. They give the silver grain to a board when quarter sawed. They carry food from the bark to the inner living portions of the tree, store reserve food material and form a means of communication with the air They are present in great numbers in all kinds of wood, thousands occuring to the square inch, though often so small as to be invisible without a strong magnifying glass. In some species as in birch and cherry their terminations in the bark are often conspicuously marked by rough spots called lenticles.

Between the bark and the sapwood is a layer of active living cells, capable of division into other cells. This is called the cambium layer. It is by the division of the cells in the cambium layer that the trunk of a tree grows in diameter and the branches and roots become larger. When first formed from the cambium layer the cells are filled with protoplasm and remain alive as long as they are in the sapwood. In young trees all the wood is sapwood. In time, varying in length with different species, this sapwood begins to change to heartwood. The cells lose the living protoplasm, become darker in color, and the cell walls harder. They no longer conduct food material to the crown. Heartwood is dead wood. It serves no purpose in the life of the tree except to give strength and rigidity to the stem. There is no definite age at which sapwood changes to heartwood. It is a more or less irregular process.

The outer layers of a tree thus form a shell of living tissue over the dead inner portions of the roots, stem and branches. Often trees are seen growing vigorously with a hollow trunk, showing that the central portion of the tree has nothing to do with the life of the tree.

DIAMETER GROWTH IN TREES

Annual Rings.—A careful examination of the bark, sapwood and heartwood will show each to be made up of layers or concentric rings, one of which is produced each year. It is possible to tell the age of a tree by counting the rings in the wood. During the spring and early summer the cells in the cambium layer divide and produce new cells. Wood cells are produced on the inside of the cambium layer and bark cells on the outside. The amount of bark produced is usually much less than the amount of wood tissue. Over the whole tree—roots, trunk and branches—is thus added each year a new layer of wood and bark.

Spring and Summer Wood.—If an annual ring of oak wood is examined it will be found to be made up of two layers, a soft, open, lighter colored layer, the spring wood, and a dense, hard, darker colored layer, the summer wood. In the spring of the year a large amount of water is needed by most trees to produce the new leaves and to supply the growing parts of the tree. As a result the cells produced in the early part of the growing season are large with thin walls. As soon as the leaves have developed, a much less amount of water is needed and the cells produced later in the season are smaller and have thicker walls. This gives rise to what are known as spring and summer wood. This difference in the texture of the spring and summer wood produces the grain in wood and separates one year's growth from another. With tropical trees this difference in the spring and summer wood does not exist. There is no rest period in the growth of trees similar to our winter, so there is continuous growth in the wood and no an-

Light affects the form of trees. Grown in the open a wide spreading crown and short thick bole results.

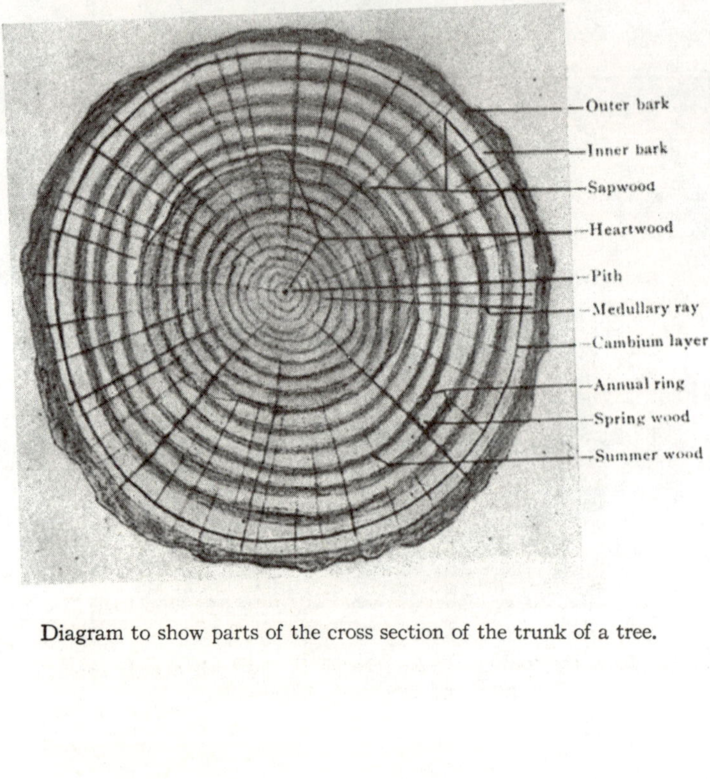

Diagram to show parts of the cross section of the trunk of a tree.

nual rings can be seen. The age of tropical trees cannot be told by the annual rings. The size of the tree however will give some indication of the age.

The width of the annual rings varies in different species. The wider the annual rings the faster has been the growth of the tree. In the same tree the width of the annual rings will usually vary from year to year. A favorable, moist year will produce a wide annual ring and a dry, unfavorable year a narrow ring. In general trees grow slowly in diameter during the first few years of their life. The rate then increases until middle life when the trees usually produce their widest annual rings. The rate of growth then falls off and in old trees the width of the annual rings may be so small as not to be seen with the naked eye.

Height Growth in Trees.—The growth of trees in height and the growth of branches in length is produced by the development in the spring of the terminal buds formed the previous season. It is a kind of stretching process or the filling out with water or sap of cells already formed in the buds. New cells are also formed by cell division and the buds quickly lengthen into twigs or shoots. These new shoots do not grow in length after the first season. The next year they will grow in thickness only. From the terminal bud formed on the end of the twig a new shoot will be produced to increase the length of the tree or branch. The growth of the roots in length and thickness is much similar to the growth of branches. It will thus be seen to be impossible for a fork in the trunk of a young tree to rise as the tree grows older. A fork once formed in a tree always remains at the same height above the ground. The trunk of a tree does not stretch. If this were not so, wire fencing nailed to a young tree would gradually be carried upward.

CHAPTER III

THE LIFE PROCESSES OF TREES AND THE NATURAL FACTORS THAT INFLUENCE THE LIFE OF TREES

SINCE a tree is a living organism it must carry on certain life processes. It must breathe, absorb and digest food, and transport nourishment to the parts needing it. Many of the life processes in trees are similar in kind to those in animals.

Breathing or Respiration.—Breathing takes place through minute openings usually on the under side of leaves called stomata. The tree breathes in oxygen and breathes out carbonic acid gas the same as an animal. The oxygen is used in carrying on the life processes of the tree. To some extent, especially in the winter after the leaves have fallen, breathing takes place through small openings or lenticles in the bark of the trunk, branches or roots, the oxygen being conducted to the inner portions of the tree along the medullary or pith rays.

Absorption and Transpiration.—The food of trees is absorbed through the root hairs and through the leaves. The food absorbed by the root hairs is water and dissolved mineral salts, that absorbed by the leaves is carbonic acid gas from the air. The water and mineral salts are carried up through the new wood to the leaves. The salts are contained in the ground in very weak solutions. In order that enough may be brought up to supply the needs of a tree, a far greater amount of water must be absorbed than is actually needed. The excess of water passes out through the stomata of the leaves as water vapor.

This process is called transpiration. The action also helps to cool the leaves as does perspiration in animals. Transpiration takes place in winter through the lenticles of the bark. The main part of the mineral substances taken up remains in the leaves and young twigs when the water is transpired and is returned to the soil when the leaves are shed or when the trees are cut and the brush left to decay. When wood is burned these mineral substances reappear as ash.

Digestion.—Carbonic acid gas forms the bulk of the food of trees. By the action of sunlight and chlorophyll, the green coloring material of leaves, carbonic acid gas is decomposed into oxygen and carbon. This process is called photosynthesis. The carbon is retained and the oxygen exhaled into the air. The carbon is then combined with the hydrogen and oxygen of the water and with certain of the mineral salts to form food substances like starch and sugar, which can be utilized by the tree. In other words, the materials absorbed through the roots and leaves must be digested in the leaves before they become available as food.

Circulation.—After food is thus prepared in the leaves it is carried down through the inner, soft bark of the branches, trunk and roots and then through the medullary rays into the wood. Some of the food material as it is carried down the tree is used at once in forming the annual rings, flowers, fruit and other parts of the tree. Other portions are stored away for starting the next year's growth. The expression "the rise and fall of sap" in trees is thus seen to have no real meaning. Water and mineral salts are constantly being taken into the tree through the root hairs and carried to the leaves through the new wood or sap wood of the trunks and roots. The process goes on in the winter time as well as in the summer although very slowly. Because of the greater transpiration from the tree in the summer due to the leaves, there is actually more water in a tree in winter than in summer. After food material has been manufactured in the leaves it is carried down through the soft or inner bark and into the tree through the pith rays. The sap never passes downward through the wood of the tree.

How Trees Reproduce Themselves.—Trees reproduce themselves principally by seeds and by sprouts. A few trees produce seed every year but most trees produce heavy seed crops only every two or more years. Seed production requires a large amount of food. This the tree stores up and after a seed year most trees require some time to lay up sufficient food to produce another crop. Trees in the open produce seed more often and in greater abundance than trees in the woodlot owing to the larger crowns. In the woodlot trees produce seed only on the upper portions of the crown which are exposed to the light. Usually a large part of the seed produced by trees is infertile.

Most trees are very particular in regard to the character of the seed bed or the ground on which the seed germinates. Some trees like hemlock and yellow birch like a moist seed bed. These trees are often found starting on rotting stumps or logs or on moss. The seed of many other trees requires the mineral soil for germination. Some, like the pitch pine, cedar and gray birch will grow on dry soil in full sunlight. Others demand shade for germination. Where the ground is too hard and dry for the tender roots from the seed to penetrate, no seedlings will be produced.

Trees vary in the arrangements for scattering the seed and the distance to which the seed is sown. Light seeded trees like ash, tulip, cottonwood and willow will scatter their seed long distances in the wind, often a mile or more in heavy winds. Many seeds have wings or other contrivances which cause the seeds to whirl about as they fall or which hold them in the air so that they are blown far from the seed trees. Trees with heavy seed like oak, hickory and chestnut can sow their seed only under their crowns, except as the falling seeds strike branches and bound away from the tree or roll down a slope.

One year old sprout from chestnut stump.

Squirrels and birds help to scatter the seeds of trees. Squirrels will carry nuts long distances from the trees that produce them and hide them in the grass or ground for winter use. Such seeds are often forgotten and germinate and grow into trees.

Usually hardwood trees reproduce themselves by sprouts as well as by seeds. When a hardwood tree is cut down sprouts start from the stump which in time grow into trees. Some trees like the chestnut will produce hundreds of sprouts from the same stump, while many other trees will produce but a few. Trees from sprouts grow faster than those from seeds during their early life, because they make use of the roots of the old tree which are stored with food. Sprout trees are not so long lived as trees from seeds and do not grow to so large a size. They are also apt to become rotten at the base because of disease communicated to them from the old rotting stumps. Trees grown from seed, while they are of slower growth than sprout trees for thirty to forty years, will finally overtake the sprout trees and outgrow them. Trees sprout best in early life. Old trees are apt to lose this sprouting ability. Trees will sprout if cut at any time of the year. If cut in summer, however, the sprouts produced will be winter killed, for the wood will not have had time to harden before the early frosts come. The time to cut trees for sprout reproduction is in winter when they are at rest. Trees growing along fences can often be killed by cutting them in summer. The sprouts produced will be frozen back the next winter. Sprouts will be sent forth the following spring but they will not be vigorous. They should be cut in summer and new sprouts obtained which in turn will be killed in winter. In this way a stump can generally be exhausted of its food and killed.

Some trees like the beech and wild plum reproduce themselves by suckers as well as by seeds and sprouts. These are shoots that start from the roots. Still another method of reproduction known as the vegetative method takes place in some trees like the willow. If a branch is broken off from the tree by the wind or by the weight of snow and ice, it will start to grow if a portion of the branch becomes covered with earth. Many trees like the willow and poplar can be reproduced by making cuttings of the new twigs and planting them in the ground. On high mountains or in the far north where the summer season is too short to mature seeds, some trees will reproduce themselves by layering. This takes place when a branch of the tree touches the ground and produces roots.

NATURAL FACTORS THAT INFLUENCE THE LIFE OF TREES

Trees cannot live without heat, light, moisture and food. These are often spoken of as factors of site. Site means the location or situation, as bottomland, side hill, upper slope or ridge. Trees vary in the amount of each of these factors demanded for their best growth. The kind of trees found in any situation is determined by the presence or absence of these factors in varying amounts. Some trees require more heat than other trees. On a moist situation are not found the same trees that grow on dry situations. Trees are seen growing beneath the shade of other trees, while others demand the full sunlight. Some trees will not grow well unless they have a fertile soil, while other trees will grow on poor soil. Trees are often found growing where one or more of these factors are not favorable to them. But they do not grow well under such conditions. They will have very slow growth and be more or less stunted and of poor form. The amount of each of these different factors needed is not definite. There is a maximum and a minimum amount of each which a certain species of tree will stand. If the limit is exceeded either way the tree will die. There is also an optimum amount for each factor under which a tree will do its best. The factors often supplement each other. On a deep, rich soil less moisture is required than on a thin, poor soil. Less light is required where the soil and moisture conditions are very favorable.

Photo by A. S. Rhodes.

White pine seeded naturally from seed sown by neighboring trees.

A "Fallen Giant." Big tree in the Mariposa Grove, California. The horsemen are a troop of U. S. Cavalry.

The action of the factors of site on tree life is very complex and it is difficult to determine often just the amount of each separate factor that is needed by a tree. Nearly all trees will grow well under favorable conditions. Only a few will live where the conditions are not favorable. It must be remembered that a site favorable to one tree may be an unfavorable site for another tree. The trees found growing naturally in the same locality and on the same site do not differ widely in their demands. For this reason forests usually are made up of a mixture of trees.

Heat.—Trees cover the greater portion of the earth from the equator to the limit of tree growth in the polar regions. Many trees grow luxuriantly in the hot climate of the Torrid Zone. Provided there is enough moisture present, the extreme heat of the sun in that region is not too great for certain kinds of trees. Some of the largest known trees of the world are found in the temperate climate, such as the redwood trees of California. But as we pass into the colder regions of the north trees which farther south grow to great height become smaller and smaller until they are shrublike and finally disappear. There is a line in the colder regions called the timber line beyond which there is no tree growth. The same gradual dwindling in size and the final disappearance of trees is noticed in ascending a mountain. Trees which grow naturally in a warm region will not grow in the colder regions of the north. Trees that grow in the temperate climate will wither and die under the hot sun of the warmer regions. There is an optimum region for each tree—a certain portion of the country where it will grow best. Much farther north or south it will not thrive and will finally be crowded out of the forest by other trees better suited to the conditions. The trees found growing naturally in any portion of the country are probably better suited to its climate than trees that grow elsewhere. There is danger in planting a tree not native to a certain region. Unless the climate where the trees are planted is similar to the climate under which the trees naturally grow, they will not thrive. Seed from trees that grow naturally in northern regions produce trees that do well when planted farther south, because they are less apt to be injured by late frosts in the spring. Southern grown trees coming from a region of longer growing season when planted farther north, do not stop their growth in the fall in time to escape the early frosts, and in the spring they are apt to start too early. There are, however, many trees not native to a region which can be introduced if found suited to the climate. Many foreign trees do well in this country when they are planted where the climate is similar to that of their native country.

Light.—The amount of light which is demanded by trees varies with different species. Some are able to thrive beneath the shade of other trees, while many demand the full sunlight and grow only in the open. The former are spoken of as tolerant trees, the latter as intolerant trees. Spruce, beech, hemlock and dogwood are examples of tolerant trees. They grow naturally in shady places as on the northern slope of a hill, in a ravine or beneath taller trees. Aspen, jack pine and tulip are examples of intolerant trees. They demand full sunlight for their proper development. If they become overtopped by other trees they will gradually die. Tolerant trees will grow in full sunlight if started in the sun. Some species like the white pine are tolerant in their youth but as the trees grow older they become intolerant and demand full sunlight. The seedlings of all trees will stand more shade than the older trees. Seedlings of many intolerant trees are often found coming up in woodlots. These will die later if the older trees are not removed, or if there are not breaks in the canopy allowing the sunlight to pass through. Trees growing under other trees if removed and planted in the sun will die, and trees that have started and grown naturally in the sunlight will die if taken up and planted in shade. Trees are very sensitive to a sudden change in the conditions under which they have lived.

Moisture.—Soil moisture is essential for tree growth. We have seen that a very large percentage of the tree by weight is water either chemically combined in the wood substance or present as free water or sap. A tree transpires moisture constantly through the leaves and through the bark in both summer and winter. For tree growth there must be present in the soil a supply of moisture adequate to supply the tree with water needed to carry on the life processes. Should the supply fall below the minimum necessary the tree will die. If transpiration from a tree takes place faster than the roots can absorb the moisture from the soil, the tree will dry out and die. This often takes place in winter when the ground freezes below the roots of the tree and excessive transpiration takes place through the bark induced by very bright sun. The death of the trees in this way is known as winter killing. In transplanting large trees it is customary to cut back the crowns so as to reduce the transpiration of moisture, until the roots have a firm hold on the ground.

The presence of moisture in varying amounts in the soil is probably the principal factor in the local distribution of trees. Soil moisture more than any other factor determines what trees will grow in a certain situation, as on a ridge, on a slope or on bottomland. Many trees demand a large amount of water in the soil and are found growing naturally in or near swamps and on damp soil, while others seek the driest situations. Some trees are very particular as to the amount of moisture in the soil, they demand a certain amount and will not grow if the amount is increased or diminished, while others will grow under different moisture conditions. The same tree is often found growing in the bottom-land and on upper slopes and ridges, but it grows best where the moisture conditions are most favorable to it. If a tree that demands dry conditions is planted on a damp soil it will soon die, and a tree that needs a moist soil will not live if deprived of that moisture or planted on a dry soil. Because a tree grows well in bottomlands or along streams is no reason to believe that it will thrive when planted on a ridge, even if the composition of the soil is the same.

Food.—Trees as well as other forms of organic life demand food for carrying on their life functions. As we have already learned trees absorb food through their leaves and through their roots. They obtain a large portion of their food from the air in the form of carbonic acid gas, the supply of which is sufficient for all trees. They obtain mineral salts and water from the soil through their smaller roots. The demand made by trees on the soil varies with different species. Some trees will grow on an almost sterile, sandy soil, while others demand as rich and fertile a soil as do agricultural crops. Pitch pine will grow on thin, dry soil on the tops of ridges, while black walnut and catalpa demand the best of soil conditions for good growth.

The canopy formed by the tree crowns.

CHAPTER IV

THE WOODLOT

A WOODLOT is a collection or stand of forest trees grown together to produce wood for use on the farm. A forest is sometimes defined as a collection of trees growing so closely together as to interfere with the natural development of the crowns. The natural form of crown in trees is observed on those growing in the open. In a park the trees stand far enough apart from one another so that the growth of the crowns is not materially interfered with. Park trees are grown for their shade and beauty. Woodlot and forest trees are grown for the production of wood. Many farm woodlots are of the park-like form due to one cause or another, but usually forest conditions prevail.

In the woodlot the trees, standing closely together, crowd each other. As they grow the sidewise development of the crowns is restricted. The energy of growth becomes concentrated in the tops of the trees. The result is quick height growth and the production of trees with small crowns but with long, cylindrical boles free of branches—the ideal form for a tree intended for use.

PARTS OF THE WOODLOT

It is possible to distinguish several parts to a woodlot—the canopy, the region of the boles, the undergrowth, the forest floor and the region of root development.

The Canopy.—The canopy is the covering of tree crowns. Within the canopy the crowns of the trees do not usually all stand at the same elevation. Some have their crowns well above neighboring crowns spreading out in the sunlight. These are called dominant trees. Other crowns occupy the lower portion of the canopy, overtopped and shaded by those above. These are the suppressed trees. Still other crowns occupy a position between these two. They have their crowns open to the light but they are narrow and restricted, due to crowding by the taller, dominant trees. These are called intermediate trees. Where the canopy is thick it prevents excessive evaporation from the soil and serves as a protection to the under portions from the drying effects of the wind and sun and from the extreme cold of winter. Forest trees will not grow well together unless the canopy is intact, so that the floor is shaded from the sunlight, bringing about and maintaining forest conditions. It keeps the air and soil moist beneath the trees and prevents a heavy growth of grass, weeds and berry bushes, that would rob the trees of moisture and food.

The Region of the Boles.—This comprises the space between the canopy and the undergrowth or forest floor. It is occupied by the boles of the trees and is high or low according to the distance the canopy is above the forest floor. The air surrounding the boles has greater humidity than the air outside the woodlot and is warmer in winter and cooler in summer than the air in the open.

The Undergrowth.—The undergrowth comprises seedlings, or reproduction as it is often called, of the trees in the woodlot, together with bushes and shrubs. When the canopy is thin or broken the undergrowth may have a beneficial effect in protecting the forest floor from the sun and from the beating effects of rain and in preventing evaporation. But if the undergrowth is too thick it may cause a reduction in the growth of trees by absorbing large amounts of mineral food and moisture from the soil. The future of the woodlot depends on the young seedling trees in the undergrowth. Cutting all the undergrowth, as is often done to open up the

The region of the boles and the undergrowth.

The forest floor. Leaves and twigs fallen from the trees form the humus.

woodlot, may cause the soil to deteriorate or may endanger the future of the woodlot. No attempt should be made to make a woodlot park-like. Forest trees grow best under wild and natural conditions.

The Forest Floor.—The forest floor consists of the ground cover of weeds and grass and the underlying mulch of more or less decomposed leaves called the humus. Where the ground cover is not thick, the weeds and grass being scattered here and there, it does but little damage and may even have a beneficial effect in helping to protect the soil and prevent evaporation when the canopy is thin. But where this growth is very thick, as is usually the case where there are but few trees in the woodlot and the canopy is broken, it will absorb moisture and nutriment from the upper layers of the soil and so lessen the growth of the trees. It will also tend to prevent new trees from starting or if any start will suppress them by their shade.

Beneath the ground cover and above the soil is an accumulation of leaves and twigs called the humus. This is usually in several layers. The upper layer is composed of leaves of the last year pressed down by the snow and flattened by rain. Underneath this is a matted layer of leaves partly decomposed. These layers together are usually called the litter. Just above the mineral soil is found a black, powdery substance, the result of the rotting of the leaves and twigs. This black powder or humus mixes with the soil and where present in considerable amounts darkens it, often making it black like muck. Humus acts as a fertilizer to the trees. It improves the soil both chemically and physically. The decay of the leaves and twigs sets free many salts which are immediately available as food for the tree. Being a vegetable substance, the humus will absorb and hold water like a sponge, giving it out gradually as needed by the trees.

The humus, or duff as it is often called, will vary in thickness according to the conditions under which it accumulates.

The decomposition of the leaves and twigs is brought about by bacteria and fungi. Air, moisture and warmth in certain amounts are essential for their action. Where one or more of these factors are not favorable to the decomposition of the leaves and twigs, humus will accumulate rapidly. In the cool north woods under a dense canopy, the humus or duff is often found over a foot in thickness. It accumulates rapidly in cold, damp situations as on a northern slope, in a ravine or beneath the dense shade of the forest. It does not accumulate in an open stand of trees where the sun has a chance to warm and dry the soil, bringing about conditions favorable for the growth of the fungi and bacteria. It is the decomposition of the leaves and twigs and the formation of the black, powdery humus and the setting free of soluble salts that is of advantage to trees—not the accumulation of the partially decayed leaves in thick layers. Too thick an accumulation of humus tends to make the soil sour because of the vegetable acids set free. Opening up the forest canopy by the removal of a few trees, allowing the sunlight to reach the floor, will bring about a rapid reduction of the humus.

The Region of Root Development.—This comprises the soil and the subsoil and the root systems of the trees. The subsoil is of more importance in forestry than in agriculture. It often determines the kind of tree that will grow. There is scarcely any soil that will not support some kind of tree growth. The depth and character are of great importance. Where the soil is not of sufficient depth to allow the development of the characteristic root form of trees, the growth of the trees will suffer. This is especially true where trees that naturally produce a tap root are grown on shallow soil. The roots of a tree extend out in all directions as far as or farther than the branches. In a woodlot the root systems of the different trees interlace, until the soil and often the subsoil down to the water table is completely filled with roots. Trees growing too closely together constantly compete with each other for moisture and food from the soil. This interlacing of the roots of trees is often of great value in preventing the soil on steep slopes from being washed away by rains or floods. A certain kind of soil may be more favorable to one kind of tree than to another. The chestnut, while it will live and develop on limestone soil is not able to compete naturally on such soil against the oaks, and other trees associated with them. These other trees are better adapted to the limestone soil and so will crowd out the chestnut.

THE DIFFERENT KINDS AND FORMS OF WOODLOTS

Farm woodlots are of many kinds and forms. These are the result of the method of treatment and the method of cutting the trees. It would be difficult to find two woodlots just alike or that have received the same treatment. All woodlots can be included, however, under a few general heads.

Kinds of Woodlots as to Species of Trees.—A woodlot has a pure stand of timber if it contains but one kind of tree. Such a woodlot would be hard to find. Nearly all woodlots contain more than one kind of tree, although there are many woodlots made up nearly of one kind. It is usual to consider a wooodlot as having a pure stand if more than 80 per cent of the trees are of one kind.

A woodlot has a mixed stand of timber if it is made up of several kinds of trees, no one of which forms more than 80 per cent of the whole stand.

Kinds of Woodlots as to the Number of Trees Present.—Woodlots are dense, open or fully stocked according to the number of trees per acre. When a woodlot has just enough trees on a given area so that the crowns of the trees meet and fully cover and shade the ground, and so there is sufficient crowding between the trees to produce rapid height growth and trees of good form, the woodlot is said to be fully stocked. If the trees are crowded so closely together that the crowns are much restricted and the growth of the trees retarded, the woodlot is said to be very dense or overcrowded. When a woodlot contains a fewer number of trees than are needed to shade the soil, the woodlot is said to be open.

Kinds of Woodlots Based on the Age of the Trees.—A woodlot in which all the trees are of approximately the same age is called an even-aged woodlot. When the woodlot contains trees of many ages mixed in together it is called a many-aged woodlot. Woodlots are also divided into mature, immature and overmature stands according to their development. Second growth timber is so called to distinguish it from the old mature trees, especially the old virgin trees, of which so few remain. Such timber where the stand is not fully stocked is apt to be very limby, due to the lower branches remaining alive and hanging on the tree a long time before being pruned off naturally.

Kinds of Woodlots Based on the Form of the Woodlot.—A woodlot is said to have a regular stand of timber if all the trees are of about the same height, the crowns of the trees forming a single canopy. Where some of the trees are of one height, their crowns forming a canopy, and the rest of the trees form a second canopy at a different height, the woodlot is called a two-storied woodlot. Such a woodlot may result from the trees being of two different ages widely separated, and also may result from the trees having different degrees of tolerance and rates of growth. An irregular stand of trees or woodlot is one in which the trees are of different heights, their crowns being mixed together, giving an irregular appearance.

Kinds of Woodlots According to the Origin of the Trees.—When the trees in the woodlot have all sprung from seed it is called a seed forest or high forest. If the trees in the woodlot come from sprouts, it is a sprout forest or coppice forest. Where the woodlot is composed of both seedling trees and sprout trees, it is a composite forest.

Seedlings and sprouts resulting from cutting out many of the trees, allowing sunlight to fall on the forest floor.

No attempt should be made to make a woodlot park like. Forest trees grow best under natural and wild conditions.

The Ideal Form of the Small Woodlot.—The ideal form of woodlot for supplying a farm with wood products would be one from which the owner could obtain at any time trees of any size or kind needed. It should furnish the farm with cordwood for fuel, sawlogs to yield lumber, fenceposts or rails for building or mending fences, or small sticks for the temporary mending of farm implements. It should contain trees having heavy, hard wood suitable for fire wood, trees having light, smooth, easily worked wood for construction purposes, and trees having durable wood for fence posts. The woodlot that would supply all these different sizes of trees at any time would have an uneven aged mixed stand of timber. The trees would be of many different kinds and of all sizes and ages.

Each of the different forms of woodlots will demand a different treatment. In order to manage them properly one must have a knowledge of how woodlots are established and how they develop, and understand the laws according to which trees grow together.

CHAPTER V

THE LIFE HISTORY OF TREES IN THE WOODLOT

THE life history of a tree growing in the woodlot among other trees differs considerably from one growing alone in the open. A tree growing by itself is not hindered in its natural development. Its crown has freedom to expand in all directions and its roots can spread at will through the soil. It lives as an individual. A tree growing in the forest, on the other hand, has not this freedom of development. Its crown is restricted by the crowns of other trees and its roots must compete for space and food with the root systems of trees growing near it. It is a member of a tree association and its life is modified by the laws governing that association. What these laws are can best be understood by tracing the development of a stand of timber from infancy to old age through the following stages:

(*a*) The seedling stage, the period before the canopy closes.

(*b*) The period of rapid height growth.

(*c*) The period of rapid diameter increase.

(*d*) The period of declining vigor in the trees.

(*a*) *The Seedling Stage.*—During the seedling stage or the earliest period in the life of a stand of timber the young trees grow as individuals. If the trees have been planted by man they are evenly spaced, possibly 5 or 6 feet apart. If the young stand has been established by seed sown from neighboring trees or from sprouts, the young seedlings will not be so evenly spaced and may in places be more or less bunched together. But in either case each little tree usually

The result of the struggle for light among chestnut trees. Long, clean, straight boles from rapid height growth and natural pruning.

has all the space it needs for the expansion of its crown and roots. As the little seedlings gradually grow in height they also spread their crowns by extending the side branches. It is not many years until the ground becomes shaded. With the coming together of the crowns the sapling stage usually begins.

(*b*) *The Period of Rapid Height Growth.*—Until the crowns of the little trees come together the soil is exposed to the sun and wind and to the beating and washing effect of the rains, and is dry and hard and less fertile than if it had been under cover. As a consequence the growth of the trees is slow during the seedling stage. As soon as the ground becomes shaded a forest floor begins to form. The soil becomes soft and light and because of the protection from the sun it contains more moisture. The leaves from the trees falling on the ground rot and begin the formation of humus, which helps to enrich the soil. Grass and weeds which tend to dry the soil disappear because of the reduced light. With the formation of the forest floor and a bettering of the soil conditions a change comes over the life of the stand, and the young trees begin to grow rapidly in height. So we see that the growing of trees together has a beneficial effect. Each tree tends to help its neighbor by protecting it from the sun and wind, and by helping to enrich the soil.

Trees growing thickly together in the forest not only have a beneficial effect on each other, they also exert a harmful effect in that they tend to modify the natural form and development of each other.

When the crowns of the trees come together further expansion in a sidewise direction practically ceases and the energy of growth becomes concentrated in the terminal buds in the tops of the trees, where alone height growth takes place. Each little tree now seems as though it were running a race with its neighbors to secure the space and light above. The contest is not an equal one. Some of the trees are naturally able to grow faster than others, because their roots are better developed, and so can secure more food and moisture from the soil, or because they have found a better soil than the trees about them, or because they are naturally hardier and stronger trees than their neighbors, and so are able to grow faster. As the trees become larger the demand for light and space increases. It is not long before the trees begin to crowd each other. Not only do they contend above for space and light, but below the ground a struggle is going on between the roots for food and moisture. Whenever a branch becomes overtopped by a branch from a neighboring tree it is weakened from loss of sunlight. The leaves on it are not able to manufacture as much food as before and do not transpire as much water. Less and less amount of mineral food from the roots is carried to them and a correspondingly greater amount is carried to the healthy leaves above. The growth of the branch gradually becomes slower, the leaves wither, and finally the branch dies. Only those branches on which the leaves are exposed to the light and are able to manufacture food can live, and as fast as these become overtopped they will also die.

Tree Crown Classes.—As the trees continue to grow those that make the fastest height growth soon outstrip their neighbors. Not only have they ample room to send their tops up into the air, but their crowns being above those of their neighbors are able to spread out sidewise. They become the most vigorous trees in the stand and are called the dominant trees. The trees that have been overtopped by the crowns of their faster growing neighbors fall behind, and their growth becomes very slow because of lack of light. These are the overtopped or suppressed trees. Those trees that are being crowded by their more vigorous neighbors, but still have a portion of their crown to the light, are called intermediate or codominant trees.

The Struggle for Existence.—The struggle that started with the coming together of the crowns soon becomes very intense. It is a struggle for existence. Not all the trees that started as seedlings can survive. As the trees continue to grow larger, a smaller and smaller number of them can occupy the space which all occupied before. Many of the weaker ones must die. Those that fall behind in the struggle are crowded out of the association. Not being able to secure the sunlight on which their life depends, they are unable to manufacture food and so die. Only those that win out in the struggle, owing to their faster growth, survive. It is a survival of the fittest. Those that live on are the survivors of hundreds, sometimes of thousands, of competitors. They win out in the struggle because they are better adapted to their surroundings than their neighbors. The struggle is a continuous one. Even the dominant trees soon come into competition with other dominant trees as they continue to grow up and spread out their crowns, and the struggle begins over again between these trees with even greater energy.

Natural Pruning.—So we see that at the time the trees are exerting a beneficial effect on each other, they are carrying on a deadly struggle. But it is to this very struggle upward for light and space that we owe the long, straight, slender trunks of the trees in the woodlot. While the trees are growing rapidly in height the lower branches are dying because of lack of sunlight. In this way the living crowns become confined to the tops of the trees. On a growing tree in the open side branches are able to expand in the sunlight and remain alive and often clothe a tree to the ground. But in the forest, as we have already seen, as soon as a branch is shaded from the sun it dies because it can no longer manufacture food. Not only do the side branches die, but the tree finally throws them off. This the tree accomplishes by compressing the dead branch next the trunk by each succeeding annual layer of wood, until finally it becomes so weakened at that point that it is broken off by the wind or snow or falls off of its own weight. This process is known as natural pruning. In this way a long, clear trunk is finally produced. The succeeding layers of annual rings laid on about the tree will cover over the wound and in a few years there will be nothing to show that a branch ever grew at that point. The branches, however, will remain concealed in the tree and will form the knots that appear when the trunk is sawed into boards.

During this period the tree grows faster in height than at any other time, and before the large pole stage is completed they will have nearly reached their total height. There is a limit to the height to which trees will grow. Some species naturally grow taller than others, but with every species there is a limit beyond which the moisture from the soil cannot be raised. When this point has been about reached there will be little additional height growth.

(*c*) *Period of Rapid Diameter Increase.*—After the trees have passed the period of most rapid height growth begins the period of rapid diameter increase. The diameter at first small has increased gradually as the trees have developed until with the completion of the rapid height growth the growth energy of the trees becomes concentrated in the production of wood material, and wide annual rings result.

Owing to the long, clean trunks that have resulted from the rapid height growth and to the shade caused by the trees growing closely together, the diameter growth put on annually during this stage is of clear wood, and the trees increase rapidly in value for lumber. It is the period of the greatest vigor in the life of the trees. Seed production becomes very large. The crowns of the trees become thicker and the ground beneath the trees is shaded more than during the other periods. Having reached their total height the tops of the crowns are no longer able to push upward, but the side branches begin to grow and strive to reach the same height as the tops. The crowns of the trees lose the more or less conical form which they have had up to this time and begin to flatten out. Because of this sidewise growth the competition between the trees

The height growth completed. The period of rapid diameter growth. Western hemlock in Washington.

The period of declining vigor. Younger trees starting under the older ones, to take their places when they fall.

continues with even greater severity than before. Each tree now strives to spread more and more of its crown to the light and many trees are crowded out and die.

(*d*) *The Period of Declining Vigor in the Trees.*—This last period is the mature stage in the life of the trees. Height growth has practically ceased and the trunks are increasing in diameter slowly. The sidewise competition of the crowns will have nearly stopped and the number of trees in the final stand will practically have been reached. As the crowns flatten out they become thinner and allow more and more sunlight to pass through to the forest floor. Grass, weeds and berry bushes begin to come in. Seedlings which are able to endure more shade than the older trees spring up beneath the trees. These little trees are the children of the forest. It is on them that the future life of the woodlot depends. When the old parent trees die and fall to the ground or are harvested, these young seedlings, which have been nourished and protected beneath their crowns, spring up and take their places, filling in the spaces left in the canopy.

The trees are ripe for the harvest. If allowed to live longer they will become attacked by insects and decay and be broken by storms. As long as trees are growing vigorously in youth they are usually able to withstand the attacks of enemies. Because of the rapid diameter growth wounds are closed over quickly and insects are not able to gain entrance. But with the slower growth in old age, wounds remain exposed a long time to the air. Fungi which cause rot in the wood gain entrance and the tree begins to decay. Owing to the weakened condition insects attack the trees and burrow deep into the wood. Storms break off the older branches which have become brittle with age. Many trees that have become weakened by disease are broken off or uprooted by the wind. Younger trees grow up to occupy the place of those that have fallen. Soon all the old trees will be gone, and a new forest, composed of younger trees, will take possession of the soil.

CHAPTER VI

ESTABLISHING A WOODLOT BY ARTIFICIAL MEANS, EITHER BY SEEDING OR PLANTING

Natural and Artificial Reproduction.—Woodlots are started both artificially and naturally. They are started artificially by planting small trees grown for the purpose or by sowing the seeds of trees. They start naturally from seed sown by the trees themselves or from sprouts from the stumps of trees that have been cut.

A field that has been abandoned will soon be covered with trees started naturally from seed blown on the field or dropped by animals or birds. Hardwood woodlots from which all the trees have been cut spring up rapidly from sprouts. It is common to see in woodlots that have not been grazed or repeatedly burned by fire, young trees that have started from seed, spring up beneath the older trees where there is sufficient light. Nearly all our woodlots have been started naturally either from seed or from sprouts.

There are but few woodlots in the country except on the Great Plains and Prairies that have been started by man. Yet artificial reproduction, as it is called, has many advantages over natural reproduction. It is a sure way of starting a woodlot, the trees will be such as the farmer wishes to grow, the woodlot will be well stocked with trees from the start, and the trees will be spaced evenly and at a proper distance apart. Its chief disadvantage is in its greater initial cost. Natural reproduction will often cost nothing. Reproduction from sprouts is obtained at no cost to the owner. Natural reproduction by seeding means sometimes the leaving of a few seed trees per acre to sow the seed, but the trees left are usually those that will be of little value for timber, and often they can be utilized later after the seed has been sown. Often natural reproduction comes from trees surrounding the area cut over. The principal objection to natural reproduction is that the trees that result will be such as nature selects. They will be the kinds that are able to get started whether they are desirable species or not. Often the better kinds of trees are crowded out because the conditions of the soil or the light may be more favorable for the poorer kinds of trees. Unless the owner by cutting out such trees determines what trees shall be present in the final stand, the result will be a product of nature, wild and irregular. Other objections are that the reproduction is apt to be more or less bunched about the seed trees or around the edges of the cut-over area, so that the tract will not be fully or evenly stocked with trees. Yet the natural method will be the one that will be used commonly by woodlot owners except where new woodlots are started, where there are no trees to sow the seed, as in the Prairie region, or where new species are wanted.

STARTING A WOODLOT BY DIRECT SOWING OF THE SEED.

The Broadcast Method.—In the broadcast method the seed gathered from trees is scattered evenly over the area to be reproduced. Usually some preparation of the soil is made before the seed is sown or otherwise the conditions must be favorable, the soil being moist and loose so that the seeds will not dry out before they start to germinate. Where preparation of the soil can be made and the seed sown can be covered in some way as by brush being drawn over the area or by being raked or harrowed into the soil, good results can be obtained. After a woodlot has been cut over and the mineral soil is more or less broken up and exposed from logs being dragged out of the woods, success will often follow broadcast seeding. The seed should be sown in the early spring. Seeding on the snow as it melts will often give good results with some species. Broadcasting after a fire when the leaves and weeds have been burned and the mineral soil exposed has often proved successful with some trees.

The Seedspot Method.—In the seedspot method the seed is sown in prepared seed spots a foot or more square. The soil is dug up and cultivated on these spots. The seed is then sown and covered well with earth. The spots are usually spaced 6 feet by 6 feet apart. Good results are usually obtained by this method. Small seeds like pine, birch and elm are very successfully planted in this way, about 20 to 30 seeds being sown in each spot. Seed producing trees having taproots like walnut, hickory, butternut, chestnut and oaks are best started by sowing the seed where the trees are to grow. With these larger seeds it is usual to plant 2 or 3 seeds in each spot. More seeds are planted than trees are wanted to allow for failures. All seeds are not fertile. Where several trees come up on a seed spot they can be thinned out and the extra trees planted where no trees have resulted. Small seeds should be covered lightly and large seeds planted about twice their diameter below the surface, and the soil well firmed above them. Seed gathered from the ground in the spring where they have been stratified naturally over winter can be successfully planted in seed spots.

Seeding in Strips or Furrows.—Instead of cultivating the entire area previous to sowing the seed, strips across the planting area several feet wide and spaced an even distance apart are sometimes prepared by plowing or otherwise breaking the surface of the soil and the seed sown broadcast on these strips. Often single furrows are turned by a plow six feet or more apart and the seed sown on the exposed soil. Brush drawn along the strips or furrows will cover smaller seeds thus sown. With walnuts, butternuts and other large seed the sods can be turned back into the furrow to cover the seed.

Large seeds are usually sown about 2 or 3 feet apart in the furrow and the furrows run about six feet apart.

Direct seeding has the advantage of being rapid and cheap. On the other hand the seed may be eaten by birds, squirrels, mice or other rodents, or it may be infertile. The small seedlings that come up the first year may be crowded out or suppressed by weeds, grass or other growth, unless they can be cultivated. Direct seeding will succeed only under favorable conditions. On poor sites and in unfavorable seasons it will not prove successful. Reproduction by seedspots, however, is a most practical method and one that can be carried out to good effect in the woodlot. Where the conditions have been favorable good results have been secured by direct seeding of walnut, butternut, ash, silver maple, red and burr oak, black cherry, and white, red, Scotch and pitch pine. In the West large areas are sometimes broadcasted with seed of the coniferous trees of the region, often with considerable success.

STARTING A WOODLOT BY PLANTING THE TREES

With most kinds of trees it is best to plant the seed in a nursery and grow the seedlings for a year or two where they can be given proper care and cultivation during the early period. Strong and hardy little trees with stocky root systems can be obtained in this way that will usually grow when planted even on poor situations.

Because it is believed large trees must be used that can be produced only at great expense, is one reason why more woodlots are not started by planting. One or two year old seedlings that can easily be raised by the farmer on the farm are the best trees to use for starting a woodlot or forest planting. It would be of advantage to every farm to have a small forest nursery for growing forest trees for starting forest plantings and for filling in bare spots in the woodlot, and for growing larger trees for planting along the road and about the house.

Such a tree nursery could well occupy a few rows in the vegetable garden where the young trees will receive care and cultivation.

COLLECTING SEEDS FOR PLANTING IN THE NURSERY

The seeds of forest trees for planting in the nursery can be purchased from seed dealers at small expense, but it is in many ways better to gather the seed from the trees when the seed is ripe and to keep it over winter for planting in the spring.

Seed can be gathered from trees in the open much easier and faster than from trees in the woodlot, which bear seed only on the upper branches. Seed from trees grown in the open will produce as good timber trees as will the seed from forest grown trees. Usually small seed must be picked from the trees before it is scattered, while heavy seed can be gathered from the ground after it has fallen. Seed should not be collected until it is ripe and should be obtained from thrifty trees of good form and without disease. Seed produced by very young trees is usually not fertile and seed from the lower branches is apt to be less fertile than the seed from higher branches. The first seeds to fall are usually not as good as those that mature later.

The time when trees ripen their seed must be known. Some, like the American elm, river birch, silver and red maple, cottonwood, poplar and willow, ripen their seeds in the spring or early summer. Seeds that mature early in the year will not retain their vitality long, so the seeds of these species should be gathered as soon as ripe and planted at once in the nursery. By Fall the seedlings will have grown considerably in height and many of them will be large enough for starting forest plantings the following spring.

Most other species of trees mature their seed in the fall of the year. Such seed could be planted at once were it not for the danger of their being destroyed by squirrels or other

Photo by Author.

Seed can be gathered from the sycamore far into the winter.

Photo by Author.

The small nursery. Growing catalpa and black locust in the garden.

Planting forest trees. One man digs the holes and the other plants the trees.

rodents during the winter or of being washed out of the soil by rain or thrown out by frost. Fortunately seed that matures late in the year will retain its vitality over winter if properly stored.

Method of Storing Tree Seeds over Winter.—Seeds like maple, ash, birch, tulip, catalpa, locust, coffee tree, and of most of the evergreen trees, will retain their vitality over winter where they are kept cool and dry. These seeds should be gathered from the trees in the fall. Many of them hang on the trees for some time after they ripen, often far into the winter. The seed should be placed in cloth bags and hung up where it will keep dry and be exposed to the outside air and also be out of reach of rodents.

The seeds of many of our trees are apt to lose their vitality if allowed to dry out over winter, such as walnut, hickory, basswood, boxelder, sycamore, oak, chestnut, butternut, osage orange and black cherry. These seeds must be kept moist over winter by stratifying them. All seeds will keep their vitality better if stratified over winter. This consists in mixing the seed with moist sand in a box or burying them in a pit in the ground. A layer of moist sand an inch or two deep is placed in a box and a layer of seed is spread over it. With large seed like the oak, hickory, walnut, etc., the layer of seed should not be more than an inch deep, and with smaller seed much less. The seed is then covered or mixed with moist sand and another layer of seed spread over the sand and so on until all the seed is stratified. Small seed can be folded in cheese cloth and thus stratified between layers of moist sand. The box should then be buried on a well drained slope, and covered with about six inches of straw or leaves and the whole covered with about six inches of earth. The box can be kept in a cold cellar if the sand is kept moist. Freezing will not injure the seed but will help to crack the shells and otherwise hasten germination. Hickory, walnut, oak and other seed with hard shells should especially be allowed to freeze. Where stratified out of doors they should be left near the surface for this purpose, the covering being made very thin. Care must be taken not to let the seeds dry out or allow squirrels or other rodents to get at them. Where there is danger of the seeds being destroyed the box or pit can be covered with wire netting.

Treatment of Seeds before Storing.—The outer coatings of seeds that are to be hung in bags over winter should be thoroughly dried so as to prevent molding. They should be spread out where they will dry in the wind but not in the sun. Seeds that are produced in pods like locust, coffeetree, and catalpa should be removed from the pods, and the outer husks should be removed from hickory nuts, walnuts and butternuts. The cones of tulip, birch, alder and evergreen trees should be gathered and dried when the seed can easily be shaken or beaten out. The fruit of osage orange, cherry and other pulpy fruits should be placed in water and allowed to ferment. The seed can then easily be removed from the fleshy covering by stirring or beating, and spread out to dry.

Treatment of Seeds Before Planting.—Seeds that are stored dry will germinate much quicker if they are allowed to soak several hours in warm water. Seeds of ash, tulip, maple and catalpa should be soaked for 3 or 4 hours in warm water and should be planted at once and not allowed to dry out. Hard seeds like coffeetree and locust should be placed in very hot water and allowed to stand for several days. The seed that swells should be planted at once and the rest given another treatment in hot water. The seed of cedar is often placed in lye for a short time to dissolve the waxy coating which prevents germination sometimes for several years.

Planting the Seed in the Nursery.—The seed should be planted in the nursery as early in the spring as possible. It is usual to plant the seed in rows, which should be far enough apart to allow for easy cultivation. Small seeds can be sown broadcast like lettuce and covered lightly with fine soil. Acorns and nuts should be sown two or three inches apart in the rows, while catalpa, ash, maple, elm and locust should be spaced not more than three-fourths of an inch apart. Seeds of low fertility should be planted thickly. The tendency in planting seeds is to plant too deep. A covering of one-fourth to one-half inch is sufficient for light seeds, three-fourths of an inch to one inch for boxelder, tulip, black cherry, ash, maple, locust and catalpa, and one and a half to two inches for chestnut, oak and other large seed. Seed should be planted deeper on light soil than on heavy soil. The soil should be made firm over the seeds so as to conserve the moisture, but should not be packed hard. Where there is danger of the surface drying out the seed bed should be thoroughly sprinkled after the seeds are planted and covered with a mulch of leaves or straw. This must be removed as soon as the seeds have germinated.

Growing Trees from Cuttings.—Willows and poplars are usually grown from cuttings. These can be made any time after the leaves fall or before the spring growth begins. They are usually made from the growth of the last year and cut from eight to twelve inches long with a slanting cut. Where made in the fall the cuttings should be bundled and buried upright in moist sand in a cold cellar until spring. The cuttings can be started where it is desired to grow the trees or they can be set out in the nursery in the spring in rows. They should be planted several inches apart in the rows and planted so that not more than one or two buds appear and the soil packed firmly around them. After growing a year in the nursery they can be set out in the woodlot. Basket willows are started in this way, the cuttings being set permanently in rows about 18 inches apart and about 8 inches apart in the rows.

Tending the Nursery.—The seedlings should be cultivated during the growing season the same as any garden crop. The weeds must be kept down and the soil frequently loosened with the cultivator or hoe to conserve moisture. During very dry seasons the nursery rows should be moistened if it is seen that the seedlings are suffering. Mulching between the rows will be found to work well in many cases. The seedlings should be protected during the first winter by mulching with straw or leaves. The seedlings of broadleaf trees will be large enough for planting the spring following the seeding.

Growing Evergreen Seedlings.—The raising of evergreen seedlings is much more difficult than the raising of seedlings of broadleaf trees. They must be shielded from the hot sun during the first year and are likely to be killed by disease. Where it is desired to grow evergreen seedlings it will be better to purchase one or two year old seedlings from nurserymen and set them out in the nursery to grow for a year or two before planting.

The seed of coniferous or evergreen trees like pines, spruces, firs, etc., are broadcasted on specially prepared seedbeds usually made 4 feet wide and 12 or more feet long. The soil is fertilized and worked with care so that it is the same degree of fineness throughout. The seed is then sown thinly over the surface at the rate of about an ounce to 5 or 6 square feet. The bed is then rolled and the seed just covered by sifting fine dirt or sand over the bed. The beds are then thoroughly soaked with water and covered with a mulch of straw or leaves or other material until the seed has germinated. The mulch is then removed and the bed covered with a lath screen that allows half shade, elevated about a foot above the beds. The great danger in growing coniferous seedlings is that when the seedlings have come out of the ground they are likely to be killed by the damping off fungus. This can be controlled only with great care and attention, by removing the screens and drying the surface soil when the fungus gets started and digging out the portions of the bed affected. As has been said, it is better to purchase one year old seedlings from nurserymen and set out in rows in the nursery for a year or two. They will cost about $2.00 or $2.50 per thousand.

Waste land should be planted with forest trees. No land on the farm should lie idle.

An even aged woodlot of white and Norway pine.

Use of Wild Stock.—Seedlings of trees for planting can often be dug up on the edges or in openings in the woods or along roads and under trees in the open. This is known as wild stock. If planted out at once the seedlings are not likely to live, because seedlings grown naturally produce very spreading root systems. When they are dug up a large portion of the roots is destroyed. If planted in the nursery for a year where they can be cultivated and cared for, new roots will be formed, making a hardy tree that can safely be planted out in the field or woodlot. Thousands of seedlings of trees are often seen growing in this way that can be collected at no cost except the labor of gathering them. It must be remembered in gathering wild stock that the conditions under which a tree grows can not be changed suddenly or the tree will die. Seedlings should not be dug from shady situations and planted in the sun.

THE TIME OF PLANTING THE SEEDLINGS

The best time to start the planting is early in the spring, as soon as the ground can be worked and before the spring rains. Hardwood trees should be planted before the buds have opened; evergreens before the new growth has started. Fall planting is sometimes practiced, especially where the moist season comes in the fall of the year instead of the spring. Seedlings planted in the fall are in danger of being heaved out of the soil by alternate freezing and thawing, and the roots not having a good hold on the ground and being unable to absorb moisture readily may be winterkilled. Planted in the spring the seedlings begin to grow at once and get a hold on the soil, and unless a very dry season follows the planting most of the trees will live. Where the planting is started by sowing the seed directly where the trees are to grow, the seed is kept over winter, as has been described, and planted early in the spring, but seed that ripens in the spring of the year should be planted at once. Cloudy days should be chosen for the work of planting. Transplanting is a shock to the young trees and all conditions must be made as favorable as possible.

Preparing the Soil.—Better results will be obtained if the planting site is thoroughly cultivated before setting the trees. This is especially true in the Middle West where the heavy grass sods will quickly kill out the little trees. Thorough cultivation of the soil conserves moisture, facilitates the planting of the trees, induces rapid growth, and increases the proportion of successful trees. Where the soil is favorable and where there is not a heavy sod of grass or cover of weeds, preparation is not always necessary. In the eastern part of the country with its moist climate it is rarely that cultivation of the soil previous to planting is necessary. It is only where the conditions are not favorable to tree growth that it would be recommended.

Spacing the Trees in the Planting.—The distance apart the trees should be planted will depend largely on the species of tree used and on the character of the site to be planted. Tolerant trees, or those that will stand considerable shade, can be planted closer together than intolerant trees. Close planting should be used where the site is unfavorable. Fast-growing trees should be planted farther apart than slow-growing trees. In general most trees should not be spaced farther apart than 5 or 6 feet. The reason for planting the trees so closely together is to get the ground covered as quickly as possible by the crowns, so that the trees will protect the soil and each other from the sun and from cold and drying winds, and bring about forest conditions rapidly. Close spacing also causes the early death of lower branches of the trees before they become large. It causes rapid height growth in trees and produces long and cylindrical trunks. Wide spacing usually results in more branchy trees with thicker trunks. Where the seedlings are planted among the stumps of trees or in brush a wider spacing will often be necessary.

SPACING TO BE FOLLOWED IN FOREST PLANTATION

From Farmers' Bulletin 711

Spacing in feet.					
12 by 12	8 by 8	7 by 7	6 by 6	5 by 5	4 by 5
Number of Trees Required per Acre					
303	680	889	1210	1743	2178
Cotton-wood.	Yellow poplar. Red gum. Loblolly pine.	Short leaf pine. Chestnut. Black locust. Bald cypress.	White pine. Red pine. Red oak. Black oak. Black walnut. White ash.	Hickory. White oak. Chestnut oak. Burr oak. Post oak. Red elm. Basswood.	Hard maple. Yellow birch. Beech. White spruce Red spruce. Fir.

How to Plant the Trees.—The seedlings should not be pulled up from the nursery, but should be lifted with a spade, so as not to strip the tender bark from the roots. They should be placed at once in pails with several inches of water in them, so that the roots will be kept moist all the time. The trees can also be placed in baskets, provided the roots are kept covered at all times with wet moss and only one seedling removed at a time for planting. When the planting site is at a distance from the nursery the seedlings can be tied loosely in bundles to facilitate handling them. The roots should be puddled in thin mud, so that they will become coated, and the bundles healed in or planted firmly in shallow trenches so that the roots are well covered, until time to ship them. In shipping, the roots should be packed firmly in wet moss, leaves, straw or moist earth, so as to exclude air and keep the roots moist. When the seedlings reach the planting site they should be healed in as before until needed for planting. Great care must be used with coniferous seedlings, for exposure of the roots to the sun and wind, even for a few minutes, will usually kill them. A seedling should not be lifted from the pail until it is ready to be planted.

There are several methods of planting seedlings. In general the seedlings should be set in the ground at about the same depth that they stood in the nursery. The soil should be pressed in against the roots as well as downward, so that no air spaces will be left, and the roots placed as nearly as possible in a natural position. The earth should be made firm about the trees.

With small trees the slit method is often used with success. This consists in thrusting a spade or grub hoe in the ground and working it back and forth so that a wedge-shaped hole is made. The roots of the seedling are then inserted back of the spade in the cleft thus formed. The spade is removed and the earth pressed with the foot firmly against the plant. This is a very rapid method and fairly cheap and successful where the soil is not too dry or poor. Care must be taken to get the roots deep down in the cleft. A folded root will usually die.

The surest method is the individual hole method. The grass and weeds are stripped off a spot about 12 to 18 inches square with the grub hoe or mattock, a hole is dug in the center and the tree planted by hand, the roots being spread out in a natural position. Long straggly or broken roots should be pruned off neatly. This is a slower and more expensive method than the slit method, but it is more successful, especially where the conditions are not very good. With larger seedlings it is the only method that can be used successfully.

Trees are sometimes planted in plowed furrows. A furrow is turned and the trees placed upright against the side. The sods can then be thrown back in the furrow and pressed against the trees by treading, or a second furrow can be made to throw the sods against the roots. This is a rapid and very cheap method, and in good soil has proved successful with many kinds of trees, such as cottonwood, maple, ash, and also with pine and spruce. It is especially applicable with poplar and willow cuttings.

The usual method of planting is for the men to work in

crews of two. One man digs the hole with a mattock and a second man follows planting the seedlings. A line of stakes is sometimes set across the field to give the line of site. Where several crews are planting together the first crew gives the line and the other crews plant the trees in rows spaced an equal distance apart. Usually no attempt is made to keep the rows exactly straight.

The Cost of Planting.—Planting usually costs from $6.00 to $12.00 an acre, depending on the cost of seedlings and of labor and the method of planting used. Two men should plant from a half acre to an acre a day, depending on the method of planting and the size of the seedlings and the character of the soil and the planting area. If the trees are raised in the farm nursery or wild stock is used, and if the trees are planted with the farm help, the cost of establishing the plantation will be very small.

CHAPTER VII

TREES TO USE IN STARTING WOODLOTS

The selection of the trees to plant in starting a woodlot or forest planting is an important consideration. It may take several years to show that trees are not adapted to a certain region or situation or soil. The ordinary planter cannot afford to experiment with trees. The loss of many years may result from a wrong choice. Trees must be adapted to the conditions under which they are planted or they will not thrive.

The following points should be considered in the selection of the trees to plant:

Is the tree hardy in the climate where it is to be planted? Many trees grow well within the natural range of their distribution. Few trees will grow well far outside of the range. This is because a tree cannot change its habits of growth to conform to a new climate. Retaining its old schedule of growth, it may leaf out too early in the spring or continue its growth too late in the fall and so be frostbitten. Trees from a region where the air is moist will not grow in a region of dry air. Trees from the Pacific coast will not thrive in the East because of this fact.

Is the tree adapted to the situation? Trees that will do well in lowlands will not thrive when planted on upland regions. Trees from high altitudes will not thrive in low land. In the same region a matter of a few feet often determines the character and kinds of trees that will grow, because of a difference in the moisture conditions. Exposure to wind and frost is an important factor in tree life. In the same region the same trees

Single row of cottonwoods, as a windbreak, closely planted, densely branchy, resulting in high efficiency.

are rarely found growing on both a steep north and southern slope.

Is the tree suited to the soil? Although soil does not affect trees to the same extent as agricultural crops, the kind of soil is important with many species. Some trees demand a moist fertile soil and will not grow well on any other kind. Whether the soil is limestone or of granite origin may determine in large measure the kinds of trees that will grow.

Will the tree selected when grown serve the purpose desired? Trees are grown in a woodlot for definite purposes, either for lumber, or fuel or fence posts or other uses. Will the form and size of the tree when grown and the character of the wood be suited for the particular use for which it is grown? Is the tree of fast enough growth to give the product desired in a reasonable time? With trees that are suited to the climate and situation, these questions become of prime importance to the woodlot planter.

Are the trees subject to attacks by insects or disease? There are certain kinds of trees that are known to be seriously damaged and often killed by attacks of insects. There are also trees that are subject to fungus diseases. Only those trees that are able to withstand such attacks should be selected.

The Use of Trees Native to a Region.—Not all the trees that grow naturally in any particular region are suited for planting in woodlots. Many of them have too slow growth, or do not grow to large enough size, or do not produce wood that is of value for use on the farm. There are, however, no better trees to plant in any region than those that grow naturally in that region, provided they will give the product desired. Such trees are suited to the climate and the conditions. No mistake can be made in their use.

In choosing a native tree to plant on a particular site no mistake can be made in following nature. The trees found growing vigorously on a site similar to the one to be planted should be selected. Trees will rarely do well on a site poorer than that on which they naturally grow. On the other hand, almost any tree will thrive under good soil and moisture conditions.

The Use of Trees from Other Regions of the Country.—There are many trees not native to a region but which grow well in some other portion of the country that can safely be planted. We often find growing around farm homes trees not native to the region that have made good growth, and that have shown that they are suited to the climate. Such examples are worthy of notice. Where such trees give the product desired they can safely be planted in the woodlot.

The Use of Foreign Trees.—There are many foreign trees that have been long in use in this country in many regions for ornamental purposes. Some of them have made good growth and seem to be adapted to the climate. A number of them have been used for forest plantings. Norway spruce, European larch, Scotch pine, and Austrian pine have been planted for this purpose. It is a question whether foreign trees will in general prove as valuable as those that are native to the country.

Trees have been planted experimentally for many years in nearly every region, and the trees suited to the climate and those that are not have been fairly well worked out. Information in regard to the trees to plant can be obtained from the State Forester or from the State Agricultural College.

ADVANTAGES OF MIXED PLANTATIONS

Mixtures are sometimes made of forest trees in planting, often with greater success than where species are planted pure.

The advantages of planting two or more kinds of trees in mixture are as follows:

1. It enables the production of more than one kind of wood. Farmers need wood for construction purposes and for fence posts, fuel, stakes, etc. There are but few species of trees that will furnish a wood suitable for all these uses.

White pine used as a windbreak.

A mixture of a coniferous tree giving a soft construction wood and a hardwood suitable for fuel and fence posts would serve the farm better than a single species.

2. Many trees have light foliage and when planted in pure stands do not shade the soil sufficiently to keep out a rank growth of grass and weeds. In this case a mixture of trees will furnish greater shade and bring about and maintain better forest conditions on the floor. Trees with light foliage are often underplanted with more tolerant trees for these purposes.

3. Some trees should be spaced widely apart in order that fast growth will be secured before the crowns of the individual trees come together to form the canopy. Larch, cottonwood, white pine and black walnut often do better with a wide spacing. In this case the final yield can be considerably increased and better forest conditions secured by filling in the spaces between the trees after they are well started with some other kind of tree.

4. Often a mixture of trees will lessen the expense of the planting. Norway pine, European larch, white pine and many other nursery grown trees are expensive. Transplants often cost from $5 to $10 or more a thousand. In planting them a cheaper tree called a filler can be used like the maple, either planting the trees in alternate rows or planting first a maple and then one of the other trees throughout each row. In this way only half the usual number of the expensive tree will be needed. The cheaper tree can be removed later in thinnings. Wild stock or seedlings that have been seeded naturally can be used for this purpose.

5. A mixed stand is less apt to be badly damaged by insect attacks or fungus diseases. With pure stands insects and diseases spread rapidly. In a mixed stand insects will reproduce less rapidly and insects and diseases will spread more slowly from tree to tree than in pure stands. Should one species be destroyed the other species, if hardy, will develop and form the future stand.

The following mixtures, taken from Bulletin 153 of the U. S. Department of Agriculture, should prove successful on soils adapted to both species:

1. Cottonwood and silver maple.
2. Cottonwood and Norway spruce.
3. Cottonwood and white spruce.
4. Cottonwood and green ash.
5. European larch and red oak.
6. European larch and white pine.
7. European larch and white spruce.
8. European larch and Norway spruce.
9. White pine and Scotch pine.
10. White pine and Norway pine.
11. White pine and hard maple.
12. White pine and red oak.
13. Black walnut and white spruce.
14. Old open stands of black walnut underplanted with white pine and of soft maple with white spruce.

Many Trees can be Widely Planted.—With the exception of the Prairie region with its scanty rainfall and the northern portions of the Lake States and the Spruce region of New England with their severe winters, the climate throughout the northern and eastern portions of the country does not differ materially. Over the whole area there is sufficient rainfall to support tree growth, and the growing season is long enough to mature many species. Many of the species found growing in the southern New England and Middle Atlantic states will also be found growing in the Central Hardwood Region. Because of these facts most of the trees found growing naturally in one region will also be found on sites suited to them in other regions or will have been widely scattered by planting. There are but few trees that will do well only in restricted regions in this part of the country. Catalpa is an example of this. It grows naturally along the bottomlands of the lower Ohio and Mississippi Rivers and is badly damaged by severe winters if planted far outside its natural range. Over this region in general the question of the selection of the trees to plant is not primarily one based on climate, but rather on soil conditions or site or on the product desired. A third factor often becomes of prime importance—that of the hardiness of the tree to insect and fungous attacks. Often the best tree to plant for a particular purpose cannot be used for these reasons. For example, there is no better tree to plant in the northern part of the region than the white pine for lumber. The white pine weevil which kills the leaders of the trees already makes the use of this tree impossible in many places. It is also likely to be killed by the blister rust should it become widespread. European larch, which furnishes a wood particularly well adapted to farm purposes, being a fairly durable wood and strong, cannot be used generally because of the larch saw fly that defoliates the trees, thus killing them. Black locust would be the best tree to plant by farmers for fence posts and fuel were it not for the locust borer which riddles the wood. Chestnut throughout its natural range has probably been the most serviceable wood to farmers, furnishing fence posts, fuel, poles and lumber. The chestnut blight prevents its being planted or favored. It is unfortunate that some of the best trees for the farm woodlot cannot be advised for general use. Trees not hardy to insect and fungous attacks should be planted only after thorough investigation of conditions.

Trees to Plant for Lumber and Fence Posts.—For the production of lumber, trees should be planted that have fast growth and a soft, light, easily-worked wood, and one that will not warp or twist excessively in use. In general the coniferous woods fulfill these conditions. Red pine, white pine, Scotch pine, European larch, Norway spruce and white spruce have been planted extensively. Douglas fir and western yellow pine are also coming into use and may prove of value. Among the hardwoods there are many trees that furnish valuable wood for lumber for one purpose or another—chestnut, white ash, yellow poplar, red oak, black walnut, American elm, and sugar maple are hardy trees and can be planted generally.

For fence posts a tree that furnishes a durable wood should be selected. Red cedar, white oak, black locust and chestnut have been valuable fence post trees. Red cedar and white oak are of too slow growth to grow for this purpose and the black locust and chestnut cannot be recommended for general planting because of the insects and diseases that attack them. Catalpa has been planted generally, but has seldom been successful. It succeeds only on the best of soil and moisture conditions. European larch is subject to the saw fly. There are but few other trees of durable woods that can be recommended. Honey locust, coffeetree and osage orange have durable woods and can be widely planted throughout the region. Almost any tree could be grown for fence posts if the wood were durable. Cottonwood is probably the fastest growing tree that could be planted, but as fence posts the wood will last but 2 or 3 years in the ground. Such wood can, however, be given a preservative treatment, as explained in a subsequent chapter. When so treated a perishable wood will last much longer usually than the most durable of woods.

FOREST PLANTING IN EASTERN REGIONS

The Spruce Region.—This region comprises the northern part of Maine, New Hampshire and Vermont, and the Adirondack and Catskill Mountains of New York and the surrounding hilly lands. Red spruce is the predominant tree, growing pure and in mixture with hardwoods and other conifers. Beech, birch and maple are the chief hardwoods, and besides the red spruce are found white and black spruce, hemlock and balsam fir. The topography is rugged. In the northern portion, because of the excellent natural reproduction of the spruce, it will be relied on to reforest cutover areas. Where extensive burns have destroyed the trees planting will be necessary. The native white spruce will do well planted in pure stands. It is easily grown in nurseries. Three year old transplants should be used spaced 6 feet by 6 feet. The Norway spruce has been used extensively for this purpose. There are many abandoned fields in the region on which it will be advisable to plant spruce for pulpwood and lumber. The southern portion of the region is hilly rather than mountainous. The soils are much better, a large percentage being devoted to agriculture. The forests are broken up into small-sized areas, there being many farm woodlots. The same species are found as in the northern portion, but in different proportions, and in addition several trees occur that creep up from the south, such as red oak, ash and basswood. Sugar maple often forms a large part of the forest on the better soils. It is often grown for the production of maple sugar. Plantings should be made on the old fields, of which there are a large number, and to fill in and restock wornout woodlots and those containing inferior species. The better kinds of hardwoods native to the region can be used for planting for cordwood and posts, and for lumber Norway pine, white pine, white spruce and Norway spruce.

The White Pine Region of the Northeast.—This comprises southeastern Maine, most of Massachusetts, northeast Connecticut and the central portions of New York and Pennsylvania. White pine is the typical tree of the region, it being in its optimum region. It is often found growing pure, especially on the poorer soils, and is usually mixed with many hardwood trees on the better soils as well as with other conifers. The region contains a large percentage of agricultural soils. In Pennsylvania there are many extensive and continuous mountain forests and much burned-over waste lands. There are many abandoned fields, and a farm is rare that does not have some land better suited for growing forest trees than for farm crops. White pine would be the natural tree for planting in the region. Owing to the white pine weevil which kills the leaders of the trees, making them branchy and bushy in form, and also because of the danger of the spreading of the white pine blister rust, a recently introduced disease, much similar in its effect to the chestnut blight, it should be planted with caution. Where it can be grown there is probably no tree that will give better results. The Norway or red pine being a hardy tree and not as susceptible to insect and fungous attacks can be used in place of the white pine. It grows nearly as fast as the white pine and its wood is much similar in character. It will also grow on drier situations. It should be planted pure spaced 6 feet by 6 feet apart or mixed with the white pine in alternate rows. Several foreign trees have been used for plantings. Scotch pine has been advised for the drier soils. Norway spruce and European larch have also been used. Many of the native hardwoods can be planted successfully where the conditions are favorable to them, especially sugar maple, basswood, white ash, red oak, black walnut, yellow poplar and cottonwood.

Southern New England and Middle Atlantic States.—This is mainly a hardwood region. It is the meeting ground of northern and southern trees, so that the forests and woodlots contain many species. Among the conifers found are white, Norway and pitch pines, red cedar and hemlock, and among the hardwoods sugar maple, birch, beech, many oaks, white ash, chestnut, hickory, yellow poplar, basswood, walnut, elm, and others of lesser importance. Woodlots are a valuable source of timber supply. In general it is a region of many markets for the products of the woodlot, and the planting and care of woodlots should be profitable. For lumber Norway pine, Scotch pine, white pine, white and European spruce, and European larch have been recommended, and also cottonwood, yellow poplar, basswood, and red oak. For fence posts black locust, where not subject to the borer, honey locust and coffeetree have durable woods and can be used. Chestnut grows well in the region and would, of course, be the best tree to plant were it not for the chestnut bark disease. European larch can also be planted where not subject to the saw fly.

The Lake States Region.—This region occupies the greater part of Michigan, Wisconsin and Minnesota. Magnificent stands of white pine formerly grew in this region, and also of Norway pine and jack pine. On the better class of soils many hardwoods are found like sugar maple, yellow birch, white ash, red oak, burr oak, basswood, poplar and paper birch. Hemlock occurs mixed with other species, and there is also some American larch and northern white cedar. There is a very large amount of cut-over and burned land in the region in need of reforestation. Probably only about 10 per cent of the whole region is under cultivation. Owing to the high value of stumpage the planting of fast-growing trees is a commercial proposition. It will be an important forest region in the future, as it has been in the past. The trees native to the region should be planted on the soils where they grow naturally; white pine will grow well on the better class of sandy soils, Norway pine on the drier soils and jack pine on the poorest and driest of the sandy soils. Scotch pine seems to thrive on most of these sites. The better class of hardwoods of the region should be used on the hardwood soils for restocking and improving the present stands in the farm woodlots. Black locust, honey locust and coffeetree can also be planted for fence posts. Catalpa has been used to some extent, but the region is too far north for this species, the severe winters killing back the new wood.

Tho Central Hardwood Region.—This region extends from central New York State and western Pennsylvania south to northern Alabama and west to Minnesota, Iowa, Nebraska and Missouri. It has in general a level topography, but with many steep slopes and deep ravines. It is typically a hardwood forest region, the forest being made up of a great number of broadleaf trees mixed in together. Red cedar is the only coniferous tree that occurs extensively. The soil and climate are well adapted to the growing of trees. Owing to the many wood-using industries in the region and the demand for wood products of all kinds, timber is of value. Many individual trees of good form and quality bring high prices for special purposes. Much planting has been done for posts and lumber. The woodlots of the region are for the most part in a culled condition, most of the valuable trees having been removed. Wornout woodlots should be restocked and plantings started on all the poorer soils of the farm. The upland forests contain oaks, hickory, sugar maple, white ash, yellow poplar, basswood, walnut, cherry and birch. The lowland forest contains elm, red maple, black ash, sycamore, willow and gums. Many of these native species are of value for planting. Black locust and catalpa have been used extensively for posts. White pine, Norway pine, yellow poplar, red oak, black walnut and white ash all should be of value in planting for lumber. White ash and green ash will grow well and furnish lumber and wood for general farm purposes and for handles.

The Prairie Region.—This region consists of three principal divisions: the Western Prairie Region, including eastern Montana, the Dakotas, south and western Minnesota and northern Iowa; the Middle West, including southeastern South Dakota, Iowa, Nebraska, Kansas, eastern Colorado and western Illinois; and the Southern Plains, including Oklahoma, southwest Kansas, northwest Texas and eastern New Mexico. This is not a region of natural tree growth except along the watercourses, owing to the lack of rainfall. Where trees are started in forest plantings, however, and cultivated until the crowns shade the soil, so that forest conditions are established, there should be no trouble in raising valuable stands of trees. There are in general two planting sites in this region, namely, the uplands, consisting of exposed rolling prairies and plains, and lowlands or bottomlands and lower valley slopes.

The Northern Prairie Region.—On the uplands, plantings

should be in the nature of wind breaks or shelter belts located along the northern and western boundaries of every 160 acre farm and two extending from north to south at intervals of 60 rods from the west side. The species recommended for the better class of soils are European larch, cottonwood, white willow, Scotch pine and western yellow pine. On the poorer soils or in very exposed situations, western yellow pine, Scotch pine, white spruce, green ash, white elm, hackberry and box-elder should thrive. Thorough cultivation is necessary until the crowns shade the soil. The trees should be closely spaced so as to bring this condition about as soon as possible. A spacing of 4 by 4 feet for conifers and 5 by 5 feet for the slowly growing hardwoods and 15 by 15 feet for cottonwood under-planted with silver maple is advised.

On the lowlands better soil and moisture conditions are found. All the species recommended for the uplands will do well, especially cottonwood, European larch, Scotch pine, silver maple, white willow, Austrian pine and white pine. On the lowlands the growing of trees for fence posts and fuel is a commercial proposition at present.

The Middle West.—Plantings will be made primarily for shelter belts and for the production of fence posts. On the better upland soils hardy catalpa, European larch, cottonwood, osage orange, white pine and white willow are recommended. For the sandy lands in the western part of the region jack pine, Scotch pine, Austrian pine and western yellow pine and red cedar are the species advised, and on the better soils honey locust, Russian mulberry, osage orange, red cedar, western yellow pine, Austrian pine, Scotch pine, green ash and white elm. On the lowlands in the region, hardy catalpa and osage orange are the chief trees for fence posts and cottonwood on short rotations for lumber, pulpwood, boxboards and staves, and for fuel alone white willow and silver maple, walnut and Kentucky coffeetree are also recommended. The spacing used in plantings should not exceed 6 by 6 feet, and on the poorer sites a closer spacing should be used. Cultivation of the soil is necessary during the first few years.

The Southern Plains.—On the uplands the southern and western sides of each quarter section should be protected by shelterbelts from the dry southwestern winds with belts running east and west every 30 to 40 rods. In the more humid eastern region cottonwood, green ash, Russian mulberry, osage orange and white elm will thrive, and black locust where it is not subject to the locust borer. In the very dry western portion of the region there are no trees that are sure of success. The following are recommended as being likely to succeed: Green ash, white elm, osage orange, red cedar, black locust, honey locust and Russian mulberry for fence posts and western yellow pine and cottonwood for lumber.

On the lowlands better conditions of soil moisture are found. The most valuable species for planting are osage orange, black locust, Russian mulberry and cottonwood for fuel and lumber. As in the other parts of this region, thorough cultivation of the soil is essential for success in starting plantings. Mixtures of a few of the species will often do better than where a single species is planted. This is because of the denser shade that is formed. Under planting the plantations with the more tolerant species will prove of value.

Planting for Windbreaks.—Windbreaks are grown for the purpose of protecting homes, cattle, orchards and fields from the wind. In many regions this is essential. There is scarcely a portion of the country that is not subject to severe winds. Windbreaks planted about fields diminish soil erosion, mechanical injury to the soil and crops and evaporation from the soil and plants.

Trees form one of the most effective means of protection against winds. Where the protecting trees are in one or two rows the term windbreak is applied. A shelter belt consists of three or more rows but not wider than twice the ultimate height of the trees. Groves and sometimes forests are also grown for protection from the wind.

The influence of a windbreak is proportional to its height and density. The influence is appreciable for a distance equal to five times the height of the trees in the windward direction and fifteen or twenty times the height to leeward. The distance increases with the velocity of the wind.

Windbreaks have a somewhat damaging effect on crops, due to the shade cast by the crowns of the trees, and also to the sapping of the soil moisture by the roots. Species having narrow crowns or light foliage and which do not extend their roots far should be selected where the injury is likely to be considerable.

Sapping the soil moisture by the roots of trees can be overcome in large measure by very deep plowing each year to cut off surface roots of the trees. Cultivating the soil near the trees will tend to induce them to secure their food closer to the trunks. Tap-rooted species like white pine, Norway pine, oaks, hickories and maples can be used. Green ash, osage orange and mulberry do not extend their roots far. Often the value of the timber and other products, as well as the protection secured, will more than repay any loss from the crops.

The trees in windbreaks should be planted closely together so that the branches will interlace and make an effective shelter. The narrower the belt the closer the trees should be planted. Trees with slender crowns can be planted closer than those with wide-spreading crowns; slow-growing trees closer than fast-growing species. In general the trees should not be planted farther than 5 or 6 feet apart. The rows are often spaced 5 to 6 feet apart and the trees in the rows spaced much closer together.

In starting a windbreak the protection is desired as soon as possible. To secure this a fast-growing tree should be selected, even though it may not be a particularly desirable tree for the purpose. Cottonwood will grow 25 feet in five or six years on good soil. Silver maple and box elder are also fast-growing trees. A windbreak can be planted using such fast-growing trees and at the same time a more valuable tree for the purpose can be planted which will take the place of the other trees in time. A single row of trees planted along the edges of fields will give good protection in many regions. The trees should be underplanted so that when they begin to shed their lower branches the smaller trees will fill in between them. The large trees can be pruned of their limbs as the lower trees increase in height. In this way the trees will furnish valuable wood product as well as protection from the wind.

In the Middle West cottonwood is the tree best suited for windbreaks when planted on good, moist situations. It should be planted in belts 125 to 150 feet wide running east and west and should not be cut until about 45 years of age, at which time the trees should be nearly 90 feet high. The belts should be underplanted a few years after the cottonwood is started with some more tolerant tree, such as green ash, red oak or the spruces. On the uplands and on the poorer situations osage orange, green ash, honey locust, Scotch pine and Austrian pine can be used.

On the northern prairies protection is necessary from the cold north winds of winter and the hot, dry, chinook winds that blow from the mountains in early spring. Windbreaks should be planted across the northern and western boundaries of each quarter section and one or two rows planted within, the trees being planted in belts from 60 to 75 feet wide. On good situations Scotch pine, Norway pine, Colorado blue spruce and the Black Hills spruce are recommended, and on the poorer situations western yellow pine, Scotch pine and Black Hills spruce.

In the Lake States white pine is advised for windbreaks, planted in belts running north and south, spaced 4 by 6 feet

Photo by R. S. Maddox.

Black locust in gulley preventing erosion.

apart. Norway spruce has also been used extensively, and white cedar is a valuable species for this purpose planted 2 by 4 feet. On very sandy soils Scotch pine and Norway pine should be used. Windbreaks are especially needed for orchard protection.

In the Eastern States white pine, white spruce, Norway spruce, white cedar and hemlock are recommended. Two rows of Norway spruce make a good windbreak when planted about 8 by 8 feet apart and the trees alternated in the rows.

Planting to Prevent Erosion.—In many regions erosion is one of the serious effects of denuding hillsides of tree growth. It not only destroys the value of the soil by stripping off the top layers and by gulleying, but the material washed away is deposited in the streams, impeding navigation, and in times of very high water and floods it may cover fertile lands with gravel and stones, making them worthless. In many regions forests should be maintained for soil protection. Tree planting on abandoned land on the farm is often a necessary operation to prevent erosion.

Gulleys formed in the soil from erosion can often be stopped by planting trees and shrubs. To do this it is necessary to hold the soil in place by some means until the trees get a firm hold on the ground. Brush, logs, stumps and stones can be used for this purpose. The head of the gulley is the important point. If the headward gnawing can be stopped the gulley can usually be controlled by planting. Brush should be piled with the tops up stream and held in place by stakes. Back of the dam formed trees and shrubs should be planted. These should be of large size, so that they will not easily be washed out of the ground. Black locust is one of the best trees to plant, as it fills the soil with a network of roots and will send up new trees from the roots. Willow is often used, and any fast-growing tree like cottonwood should prove valuable. Shrubs planted along the edges of the gulley will prevent its growing larger.

Raising Christmas Trees.—The raising of Christmas trees is a profitable undertaking in many regions. Trees from 6 to 8 feet high are in greatest demand, and there is a good market for smaller trees. Especially well-formed specimens are often sought by nurserymen for ornamental purposes. In from 5 to 6 years after planting the trees an income should be secured. There are many conifers that can be used for the purpose. Spruce and fir are usually in greatest demand, though many of the pines like white pine and Norway pine are sold. On fairly good soil these trees should grow from a half foot to a foot each year. Because the trees are to be removed early in their life a close spacing can be used. Three by three feet is often used for spruce and fir requiring 4840 trees to the acre. Faster-growing trees should be spaced 4 by 4 feet, requiring 2723 trees to the acre. Small trees for planting can be purchased from nurserymen often for as low a price as $2 to $3 a thousand.

CHAPTER VIII

TREES AND THEIR VALUE FOR PLANTING

White Pine (*Pinus strobus*).—The white pine is widely distributed throughout the northeastern part of the country, extending south to Pennsylvania and Iowa and along the Appalachian Mountains to Georgia. It grows to large size, often living 450 years and reaching 175 feet in height and 3 to 5 feet in diameter. It grows best on deep, light, moist and sandy soils. It will, however, grow on nearly all situations where there is sufficient moisture, from steep rocky slopes to river bottoms that are well drained. The wood of white pine is soft, light, straight grained and easily worked and will not warp. It has been an important construction timber. More lumber has been cut from white pine than from any other species, and it was formerly used for more purposes than any other wood. It is gradually becoming scarce and the better grades are high in price.

The seedlings of white pine for planting, as with other conifers, are difficult to grow without constant care and attention during the first few weeks. It is best to purchase 2-year-old seedlings from a commercial nursery. They can often be purchased for as low as $2 or $3 a thousand. They will be large enough to plant on favorable sites and for under-planting open stands of timber. Where the conditions are not favorable, as where the grass or weeds grow high, the seedlings should be grown in the farm nursery for a year or more before planting. In starting plantings the seedlings are planted in rows 6 to 8 feet apart and spaced the same distance apart in the rows, the distance depending on the conditions of the soil and the situation. On the poorer sites the trees should be planted closely together.

White pine grows in height at an average rate of about a foot a year. Often a growth of several feet is made a year on good soil. At 25 years of age under good conditions white pine will vary in height from 25 to 30 feet, and in diameter from 5 to 6 inches. The trees are badly damaged by the white pine weevil, an insect that bores into the tip or leader and kills it. This may cause a bushy or forked tree. The trees are also subject to several diseases, especially to the blister rust. For these reasons it must be planted with caution. Although one of the fast-growing conifers and producing a wood more valuable for many uses than any other tree, it may be necessary to discontinue its use for forest planting because of these enemies. It is common to mix the white pine with some other tree, so that if the white pine fails the other tree will form the future stand, a row of white pine alternating with the other species. Norway pine is a good tree for this purpose, as it grows nearly as fast as the white pine. European spruce, European larch, red oak, white ash and hard maple are other trees that can be used.

Norway or Red Pine (*Pinus resinosa*).—The natural range of the Norway or red pine is along the northern border of the country west to Minnesota and south to Pennsylvania. It grows to be a tree 70 to 90 feet high and 2 to 3 feet in diameter. It grows best on a sandy loam soil, well drained, and of fair fertility, but it also thrives on poor, dry, sandy soils. It is a hardy tree and not so susceptible to insect and fungous attacks as is the white pine. It is propagated the same as for white pine. It is often planted mixed with the white pine. It grows well in pure stands. The wood is much like white pine, though a little heavier and harder. At 30 years of age trees will reach 30 to 35 feet in height and from 5 to 6 inches in diameter. Being a native tree and not injured seriously by insects or disease, Norway pine should hold an important place in forestry in the future. It is propagated as for white pine.

Jack Pine (*Pinus divaricata*).—Jack pine is a northern tree extending south to northern New England, New York, and the Lake States. It is most abundant on the sand plains of Michigan, Wisconsin and Minnesota. It is a comparatively small tree. Mature stands will not average more than 70 feet high and 10 inches in diameter. Individual trees are found 100 feet high and 18 inches in diameter. Its importance lies in the fact that it will grow well on poor, dry sands. It forms a tap root that grows deep into the ground, making it possible for the tree to stand very dry conditions. It is a very intolerant tree. It makes very rapid growth in youth. At 10 years of age trees on poor soil are often 18 feet high. It will sometimes grow 3 feet a year at that age. Its growth slows down later, so that at 60 years of age the trees are only 60 feet high and from 5 to 8 inches in diameter. It has short life, the average age of stands seldom exceeding 80 to 90 years. It bears large quantities of seed every year. Cones are often produced on trees at 5 years of age. The seeds are light and are blown long distances by the wind. It quickly seeds up burned over areas. The cones, which naturally open with difficulty, are dried out by fire and open readily after it has passed to scatter the seed. It can easily be propagated artificially by sowing the seed. The wood is soft, light and brittle, but has many uses, including bed slats, boxes, ceiling, crates, flooring, furniture and siding. It will be an important tree locally in its natural range on soil too dry and sandy for other species of trees to grow.

Scotch Pine (*Pinus sylvestris*).—Scotch pine is a native of Europe and has been planted extensively in this country chiefly for ornamental purposes. Many plantations of this tree have been started. It grows to be a tree 70 feet high and 1½ to 3 feet in diameter, and may reach 120 feet in height and

3 to 5 feet in diameter. It makes rapid height growth for the first 15 or 20 years, often 20 to 30 inches a year. It will grow on all classes of soils, even dry sand, although it prefers deep, well drained, sandy loam. It is very intolerant of shade. The wood is comparatively soft, light, resinous, and light reddish-brown in color. It is used for general construction, lumber, railroad ties and poles. Two-year-old seedlings can be purchased from nurserymen and planted in the farm nursery for a year or two before being planted, except on very favorable sites where the grass and weeds do not grow high, when the two-year-old seedlings can be used. It should be spaced about 6 by 8 feet, being rather intolerant. It is often mixed with white pine in alternate rows. It is a very hardy tree and free from diseases. It grows well on the sandy soils of the Lake States, in the Prairie region, and on abandoned lands in New England. Trees planted in this country have usually come from a mountain variety and do not grow to large size. At about 20 years of age they become crooked and yield but little timber. Unless a better variety of seed can be secured some of our native species will prove better in the long run.

White Spruce (*Picea canadensis*).—White spruce occurs naturally throughout the northern tier of states to Montana and crosses the continent and reaches the Pacific coast in Alaska. It is a tree of the cold North. It is a tree of medium size, a large tree of the species being 100 feet high and 30 inches in diameter. The wood is soft, light, and not very strong. Like other spruces it is used for the manufacture of paper pulp and for lumber and construction purposes. It will grow under a wide range of soil and moisture conditions. It is found on low land along streams and on the sides of ridges and hills. It grows best in moist, well drained, sandy loam soil. It is a very tolerant tree, the side branches remaining on the trees many years producing knotty lumber, unless grown in dense stands. Being a native tree and adapted to many soils and situations, it is believed that the white spruce will be an important tree in the future for planting for pulp wood and lumber. In many places it is being planted for this purpose. It is often used for windbreaks. It is readily grown from seed in nurseries and easily handled. After the first few years its growth is fairly rapid. It should be planted in crowded stands, spaced about 4 feet by 4 feet apart. As with the other conifers small seedlings can be purchased from nurserymen and grown for a year or until large enough to set out. On the more favorable sites 3-year-old transplants will be large enough.

Norway Spruce (*Picea excelsa*).—This tree is a native of Europe and Asia, but is hardy and grows well in the eastern part of the United States. It has been planted extensively for ornamental purposes and is often seen growing around farm homes and in cemeteries and for windbreaks. In Europe it grows to be a very large tree, often exceeding 125 feet in height and 3 feet in diameter. It grows well on nearly all situations, except a naturally dry soil. It will grow well in a fresh shallow soil, whether fertile or not, and will thrive with moderate amount of moisture. Because of its dense crown it has been planted extensively for windbreaks in the Middle West. The wood is light, soft, splits easily and does not warp. It makes a good construction material, and is used for paper pulp, masts, oars, and many other purposes. Two-year-old seedlings can be purchased from nurserymen and planted for a year or two in the farm nursery before planting. For lumber the trees should be spaced about 4 or 5 feet apart. For windbreaks a wider spacing is better, often up to 12 feet by 12 feet. It is a good tree for underplanting old groves of cottonwood, walnut or other trees where the canopy is thin, being a tolerant tree. Spruce trees grow slowly for the first few years, but after that period is passed growth is fairly rapid. Trees 30 years of age have often reached 50 to 60 feet in height and 6 to 8 inches in diameter. Because of its being a foreign tree it is probable that the native white spruce will in the end be a better tree to plant. Trees at about 40 or 50 years of age begin to grow ragged in the tops, showing that the tree is not especially adapted to the climate.

European Larch (*Larix Europea*).—European larch is a native of the northern part of Europe, growing in the cooler and higher altitudes. It has been planted extensively in this country for ornamental purposes, and also for posts and poles. It is a fast grower, but seldom reaches more than a height of 80 feet and a diameter of 2 to 3 feet. At 40 years of age trees often reach 50 to 60 feet in height and 6 to 7 inches in diameter. It is an intolerant tree, and so does not cast a heavy shade on the ground or maintain good forest conditions. It should be mixed with some other tree that will furnish the necessary shade, as white pine, white spruce or red oak. The larch should be spaced about 10 feet by 10 feet and filled in with the other species. It requires a fresh, well drained, moderately heavy soil. It does not do well on sand or gravel or on wet situations. The wood is heavy, fairly hard, and durable in contact with the ground. It would be a good tree to plant for fence posts were it not for the danger of the larch sawfly, which attacks and kills the trees by defoliating them. Seedlings for planting can be purchased from nurserymen.

Red Cedar (*Juniperus virginiana*).—The red cedar is the most widely distributed coniferous tree in the United States, being distributed throughout the forests east of the Dakotas and Texas. In the region of its best development it forms pure forests, but it usually grows in association with a large number of hardwood trees. It grows under almost any condition of soil and moisture, from dry rocky ridges to fertile bottomlands. It is particularly well adapted to the semi-arid plains, where but few other trees will succeed. In the South it sometimes grows 100 feet high and 3 feet in diameter, but usually it does not exceed 30 feet in height. The growth is naturally slow, but the rate varies with the conditions of the soil. It takes from 1 to 15 years to grow an inch in diameter. The height growth does not exceed a foot a year. The wood is light, soft and of compact grain. It will take a high polish. The wood is very durable, which makes it a valuable tree for fence posts, poles, chests and railroad ties. The wood is used in the manufacture of lead pencils. Its dense crown makes it a good tree to plant for windbreaks on the Plains. It is reproduced only by seed. Usually it takes two years for germination. The seed should be collected when ripe, soaked in warm water for several hours and stratified in moist sand and planted in the fall, a year from the time they were gathered, or kept until the following spring. They are difficult to grow, owing to the damping off fungus. It will be better to purchase small seedlings from nurserymen and set them out for a year or two before planting. It should be planted with light demanding trees like cottonwood and green ash. It will grow more slowly than these trees and form an understory shading the ground from the sun. When planted in pure plantations on dry uplands it should be spaced 4 by 4 feet apart. On better or more moist soils 4 by 6 or 8 feet is best. It is often injured by fungus diseases and by insects.

American or White Elm (*Ulmus Americana*).—The American elm is distributed widely throughout the eastern part of the country east of the Rocky Mountains. It seldom forms a large percentage of the trees in the forest. Mature trees vary from 60 to 120 feet in height and from 2 to 8 feet in diameter. It grows best on deep alluvial soil with a constant supply of moisture. It will, however, grow on fairly dry soil. It is considered one of the hardiest trees for Prairie planting, as it is able to endure the greatest extremes of temperature and drought of the treeless west. It is grown throughout its range as an ornamental and shade tree. The wood is moderately strong, coarse grained, difficult to split, not durable and liable to warp and check in drying. In young trees the wood is very tough. It is used chiefly in the manufacture of agricultural implements and carriages and for cooperage and saddletrees. It is propagated by seeds. These ripen in May and should be collected from the ground and at once planted in the nursery in drills 8 to 10 inches apart. The seed should not be allowed to dry out. The young seedlings can be transplanted when a year old. It can be grown in pure stands or in mixture with hardy catalpa, ashes, locusts, coffeetree, black walnut, black cherry, red cedar and European larch. It should be planted 4 to 6 feet apart. The tree is seriously affected by insects, especially the elm leaf beetle and other leaf destroyers and by borers that girdle the inner bark, killing the trees. In many regions it can be grown only by constantly combatting such insect pests.

Coffeetree (*Gymnocladus dioicus*).—The coffeetree is widely distributed throughout the eastern part of the country. It is nowhere abundant, occurring only as solitary individuals among other hardwoods. Forest-grown trees make a height of 60 to 100 feet and a diameter of 1½ to 2 feet. Under very favorable conditions the height growth for the first 30 to 40 years often amounts to 1 to 2 feet a year, while the annual diameter growth is about ¼ to ½ inch a year. On poor soil it is a slow-growing but long-living tree of small size. The wood is heavy, moderately hard, very stiff, and of coarse texture, and durable in contact with the soil. It shrinks and checks considerably in drying, but works and stands well and takes a high polish. It is used for cabinet work, rough lumber and fence posts. It grows best on the richest bottomlands and in moist ravines. It will grow, however, on the dry sandy and gravelly soils of uplands. It is a hardy tree and will stand the cold winters of Minnesota and the hot summers of Oklahoma without injury. The comparatively large seed is produced in a pod which hangs on the tree often until far into the winter. They should be collected and the seeds removed and stored in a cool, dry place. Before planting in the spring they should be placed in a pail of hot water and stirred for 15 minutes or more and then covered and allowed to stand 2 or 3 days. The seed that swells should be planted at once and not allowed to dry out and the rest treated with hot water. The seed should be planted in the nursery about 2 or 3 inches apart in rows 2 feet apart. They will be large enough when 1 year old to transplant. It should be planted in mixture with white elm, hackberry, oaks and ashes. If planted several years previous to its associates it can be mixed with catalpa, Russian mulberry or black locust. It is not known to be subject to any serious insect or fungus diseases. It should prove a valuable tree to plant for fence posts.

Cottonwood (*Populus deltoides*).—The cottonwood grows naturally throughout the Eastern portion of the country. It grows best on alluvial soils along watercourses. It demands moisture, but does not demand a rich soil. It grows to be a tree 100 feet in height and 2 to 3 feet in diameter. The wood is soft and light and difficult to split when dry. It is not strong and decays rapidly in the ground. It can be used for construction purposes, where properly piled after sawing to prevent warping. It is also used for packing boxes, wood pulp and crates. It will grow fence posts in ten years, trees at that age often reaching 6 to 8 inches in diameter breast height and 30 to 35 feet in height. The wood is not durable, but when treated with creosote, posts will last at least twenty years. Cottonwood has the fastest growth of the trees that are commonly planted. Where the soil is suited to it trees 20 years of age will reach a height of 50 to 60 feet and a diameter of 12 to 15 inches. A variety called the Norway poplar is being used extensively for planting purposes. Cottonwood, like willow, will start naturally from cuttings. These should be made about 8 to 12 inches in length from the last year's growth. Seedlings of cottonwood can also be found growing wild along streams. These can easily be transplanted. The trees in plantations should be spaced about 12 by 12 feet apart. It is best to underplant with some such species as sugar maple. This will insure good forest conditions. The cottonwood will produce lumber and the maple cordwood.

Basswood (*Tilia Americana*).—The basswood or linden tree is distributed from Canada to Alabama along the mountains and west to Minnesota and Texas. It grows to be a tree 70 to 80 feet in height and 2 feet in diameter. It is suited to deep, rich, bottomland soils, but will grow slowly on poor upland soils. It is a hardy tree and moderately tolerant of shade. It makes a fairly rapid growth. It is often subject to attacks of insects and fungi. The wood is soft, light brown in color, straight grained and easily worked, but is not very durable. It is used for lumber, woodenware, carriage boxes, trunks and paper pulp. Its flowers yield large amounts of fine honey. It is reproduced by both seeds and sprouts. The seed ripens in September or early October. It should be collected and either planted at once or stratified in moist sand over winter. It is a prolific sprouter, one of the best. Seedlings raised in the nursery can be set out when one year of age. They should be spaced about 5 or 6 feet apart. It will do fairly well in pure stands and also in mixture with white or Norway pine, white ash, elm, maples and hickories.

Black Cherry (*Prunus serotina*).—This tree is common throughout the eastern half of the United States. It grows to be a tree 80 feet in height and 4 feet in diameter, but is usually found about 60 to 70 feet high and 2 feet in diameter. It will grow fairly well on dry situations, but on moist, well drained, fertile soils it reaches its maximum development. It grows well on well-drained bottomlands. Under favorable conditions it is a rapidly growing tree. A growth of 25 feet in height and a diameter breast height of 6 inches has been obtained in ten years on very rich soil, but usually the growth is much slower. The wood has a beautiful grain and color and is in demand for making furniture. The heartwood makes a good wood for fence posts, it being very durable.

The sapwood rots quickly. A small tree, being all sap, would have to be treated with creosote to use for fence posts. The cherries should be gathered in August or September, the pulp removed by crushing or by mascerating in water, and the pits stratified in moist sand over winter. The seed should be planted in shallow drills in the spring about 2 inches apart. The following spring the seedlings can be set out in the field. The tree should be mixed with some other species, such as white ash, black walnut, sugar maple, etc., rather than be planted pure. At 20 years of age under favorable conditions black cherry will reach a height of 25 to 30 feet and a diameter of 5 to 6 inches.

Black Locust, Yellow Locust (*Robinia pseudacacia*).—The natural range of the black locust is in the Eastern United States along the Appalachian Mountains, but it has been planted extensively all over the Eastern part of the country. It does best on deep, well drained, fertile loam, but will grow on almost any soil except a wet, heavy one. It grows exceedingly well on limestone soil. It is a rapid-growing tree. On good soil it will grow from 2 to 4 feet in height a year. The wood is strong and very durable. It is well adapted to fence posts, ties, insulator pins, etc. Fence posts can be grown in 20 years. At that age trees will average 5 to 6 inches in diameter at breast height with a height of 40 to 50 feet. The great danger in raising black locust trees is that they are likely to be destroyed by the locust borer, which riddles the wood of the trunk and branches. This insect attacks trees of all sizes, and there seems to be no practical treatment or remedy. It is claimed that the borer lives on the pollen of the goldenrod, and when this does not run wild black locust can be grown. Before starting a plantation of this species the owner should determine whether or not the locust borer is injurious to locust trees in his vicinity. The pods of locust can be collected from the tree when dry, and hung in bags over winter in the barn. Before planting the seed should be removed from the pods and soaked in hot water. The seed should be planted in rows in the farm nursery and the little trees cultivated for a year. The following spring they will be large enough to plant out in the field. Where it is impossible to grow the black locust on account of the borer, the honey or thorny locust can often be grown.

Black Walnut (*Juglana nigra*).—Black walnut grows well in the Eastern portion of the country, especially throughout the Central Hardwood Region. It grows to be a tree 110 feet in height and 4 to 6 feet in diameter. Usually trees grow from 70 to 90 feet in height and 2 to 2½ feet in diameter. It attains its best development in deep, rich bottomlands or on fertile hillsides. It demands a moist, fertile soil. Black walnut is not a very rapid grower, but in good soil the growth is fairly rapid. It will take 60 to 80 years to grow to lumber size. It begins to bear fruit at 12 to 15 years. Trees 20 years of age on good soil have often reached a diameter growth of 5 to 6 inches and a height growth of 20 to 25 feet. Under the best conditions it often grows an inch in diameter every year. The wood is heavy, hard, strong and of coarse texture. It shrinks and warps moderately in drying. It has many uses. The heartwood is very durable in contact with the soil. It makes a fine fence post. Small trees, however, are all sapwood and should be creosoted when used as fence-posts.

For raising seedlings to plant the walnuts should be gathered in the fall, freed from the outer husk and stratified in moist sand over winter. The seed should be planted where it is desired to have the trees grow, because of the deep tap root produced. To prevent destruction by rodents the seed should be treated with red lead before planting; 2 or 3 seeds should be planted in each spot in the field so as to insure a full stand. The seedlings can be thinned out after the first year. The trees in the plantation should be spaced 6 to 8 feet apart. Being an intolerant tree, older plantations should be underplanted with a more tolerant tree, like white pine or Norway spruce or red oak, to maintain forest conditions.

Box Elder (*Acer negundo*).—Box elder is distributed generally throughout the eastern part of the country, occurring as scattered trees or groups among other hardwoods. It occurs naturally in the deep, moist, soil of valleys and on the borders of lakes and swamps, but it will grow on deep, sandy soil, and will bear exposure and thrive on rocky slopes and on the semi-arid prairies. It is a small tree, rarely exceeding 30 to 40 feet in height and 1 to 2 feet in diameter. It is usually a rapidly growing tree, but its growth depends largely on the character of the soil. The wood is soft, weak, light, close grained, and creamy white in color. Its principal use is for firewood, and it is occasionally used for interior finish, woodenware, cheap furniture and paper pulp. Maple sugar is sometimes made from the sap. The seed is produced in great abundance. They should be stratified in moist sand over winter, and planted in the nursery an inch or more apart in rows 2 to 3 feet apart and covered with an inch of soil. The seedlings often grow 10 to 14 inches high the first season. In plantations the trees should be planted closely, about 4 by 8 feet apart. It is often mixed with white elm, European larch, green ash and black walnut. It is often planted for windbreaks because of its fast growth. Because of its brittle wood the branches are often badly broken by ice and snow.

Chestnut (*Castanea dentata*).—The chestnut is distributed from Maine to Michigan, south to Delaware and along the mountains to Alabama, Mississippi and Arkansas. It reaches a height of 120 feet and a diameter of 13 feet or more. Usually at maturity the trees are 60 to 80 feet high and 3 to 4 feet in diameter. It occurs singly or in groups or groves, and seldom forms pure stands naturally over wide areas. It has a more rapid growth than any of the more valuable hardwood trees. It grows in height from 15 to 20 inches a year for 20 to 30 years. Sprouts grow much more rapidly, often 4 to 6 feet the first year, and in 20 to 25 years they are large enough for poles and ties. After 30 years the rapid growth ceases. In the end the trees from seed will reach a much larger size. The rate of growth will vary with the nature of the soil. The wood is light, fairly soft and rather coarse grained. It splits and works easily, checks and shrinks considerably in drying, and is very durable in contact with the ground. It is used in cabinet work, cooperage, interior finish fence posts, telegraph and telephone poles and crossties. It is one of the most useful of woods in its natural region of growth. It is propagated by seeds and by sprouts from the stump. There is probably no tree that sprouts better. It is easily managed by the coppice system. The nuts should be gathered in the fall and stratified in moist sand over winter, and planted in the nursery a foot apart in rows that are 3 feet apart and covered 1 to 2 inches deep. In one season the seedlings will be a foot or more high and can be transplanted when one or two years old. It is better to start the trees where the trees are to be grown. The chestnut would be one of the best trees to plant for lumber and fence posts if it were not for the chestnut bark disease, which has killed the trees over a wide area in the eastern portion of its range. Until the extent of this disease is known and its spread ceases it cannot be advised. It should be planted with caution.

Green Ash (*Fraxinus lanceolata*).—The green ash is distributed over the greater part of the United States east of the Rocky Mountains, occurring singly or in small groups in the forest. It is a medium-sized tree. A height of 80 feet and a diameter of 24 inches is rarely exceeded. It grows best on low, moist soil, but will grow under a wide variety of soil and moisture conditions. It has the ability to thrive under adverse conditions. It is one of the best trees for planting on the arid plains of the Middle West. It is not a rapid grower, but under average conditions will grow to fence post size in 15 to 20 years. It grows slowly in arid or semi-arid regions.

The wood is hard, heavy, strong, and rather coarse grained. It is used for the manufacture of agricultural implements, carriages, furniture, and often as a substitute for white ash. It has a relatively high fuel value. Although the wood is not very durable, it is often used for fence posts, especially where other species are rare. It is reproduced by seed and by coppice growth. The seed should be collected from the trees and either kept dry over winter or stratified in moist sand. If kept dry the seed should be soaked in water several hours before planting. It can be propagated by sowing the seed broadcast. The trees should be planted closely together, 4 by 4 feet, because the tree does not have a heavy foliage. It is often mixed with hackberry, box elder, white elm, Scotch pine and red cedar. It is sometimes planted for windbreaks.

Hackberry (*Celtis occidentalis*).—The hackberry is a tree of wide distribution from Massachusetts to Oregon and from Canada to New Mexico and Florida. It is found in the forest in mixture with other species. In rich soil it grows to be a tree 100 feet in height and 3 feet in diameter, but on ordinary soil it is a much smaller tree. It is a long-lived tree and of slow growth. It will live on almost any kind of soil and with a very small amount of moisture. It will live where almost any other tree would die. It is adapted for planting in the arid regions of the Middle West. The wood is straight grained, light and elastic. It is used chiefly for fuel and the manufacture of cheap furniture. It is not durable in contact with the soil and has little strength. It is one of the best trees to plant for windbreaks on land too dry for cottonwood in Minnesota, North and South Dakota and northern Nebraska. For this purpose it should be planted in double rows 3 feet apart with the trees 4 feet apart in the rows. It is often planted with light-foliaged trees like green ash and cottonwood to shade the soil. It is a tree adapted to mixed stands. It seeds abundantly. The seed should be picked from the trees in the fall of the year and stratified in moist sand. It should be planted in the nursery in drills about an inch deep and allowed to grow for 2 years before being transplanted. It has few insect enemies or diseases.

Hardy Catalpa (*Catalpa speciosa*).—The natural range of the hardy catalpa is a limited region in the Ohio and Mississippi valleys from Indiana to Arkansas. It has been widely distributed by planting. It is subject to injury by severe winters far outside of its natural range, which kills back the last year's growth. Its growth depends on the character of the soil. It will succeed only on moist, well drained, deep, fertile soil. When planted under other conditions it will not thrive. It is not adapted to stiff clay soils nor to dry conditions. It has been widely planted by farmers throughout the east for fence posts, but with success only on most favorable soils. It is not a tree for general planting. It succeeds well in the Middle West where the conditions are adapted to it, and often makes extremely rapid growth. A height growth of 2½ feet a year and a diameter growth of ½ an inch a year are not uncommon for the first 15 or 20 years. It is a valuable fence post tree in many regions. It matures early and makes a small to medium-sized tree. Owing to the fact that it does not produce a terminal bud, it forms a very crooked tree, and can be grown to fence-post size only by constant pruning. The wood is soft, light, elastic, has fairly high fuel value, and is very durable in contact with the ground. It often lasts 30 to 40 years as fence posts. It reproduces freely by seeds and by sprouts. The pods should be collected in the late fall or winter and the seed hung in a dry, cool place. Before planting the seed should be soaked 24 hours in water. The seed has high germination per cent and is an easy tree to grow in the nursery. The seedings should be planted about 6 feet by 6 feet apart. After growing a few years the trees are sometimes cut close to the ground and a single sprout allowed to develop. This will often grow 8 to 10 feet the first season. The tree is not usually subject to diseases, but is often attacked by insects.

Honey Locust (*Gleditsia triacanthos*).—The natural range of honey locust is from western New York and Pennsylvania south to Georgia and west to eastern South Dakota and Texas. It does not form pure stands, but is distributed throughout the hardwood forest. It grows to be a tree 140 feet in height and 6 feet in diameter, but usually is about 40 to 50 feet high and 1 to 2 feet in diameter. The growth is fairly rapid. On favorable sites it makes an annual height growth of 1 to 2 feet and an annual diameter growth of one-third to one-half inch a year. Under less favorable conditions it takes 7 to 8 years to grow an inch in diameter. It is intolerant of shade. It naturally produces large clusters of thorns along the trunk and branches, but these can be removed. There is a thornless variety of this species. Often in a grove of honey locust individuals will be found without them. The seeds of these trees can be collected and grown. Those that produce seedlings with thorns can be thrown away and the thornless specimens planted. The wood is heavy, hard, strong and coarse grained, and fairly durable in contact with the ground. It is used chiefly for fuel, fence posts and poles. It is useful for hedges and windbreaks. The soils of the Plains and Prairie region are well adapted to this tree. It will withstand very dry conditions. It seeds abundantly and is easily propagated. The pods should be collected from the ground and the seeds removed and stored in a cool, dry place over winter, and treated with hot water before planting in the spring. The swollen seeds should be removed and planted at once and the operation repeated with the remainder. The seeds should be sown in moist, rich soil in the nursery and covered lightly. They will grow a foot high the first year and will be large enough to transplant to permanent situations the following spring. It should be planted closely together. Two by eight feet is recommended for Prairie regions. For shelter belts it should be planted 4 by 8 feet. It is often underplanted with a more tolerant species of tree. The honey locust is not so generally injured by the locust borer as is the black locust. It should prove a valuable tree to plant for fence posts in many regions.

Osage Orange (*Toxylon pomiferon*).—The natural range of the osage orange is from the Arkansas River south through southeastern Indian Territory to southern Texas. It has been widely distributed by planting. It is sometimes called the mock orange tree because of the shape and appearance of the fruit. It is a very hardy tree. It adapts itself to a wide range of soil and moisture conditions, and its ability to endure drought makes it one of the best trees to plant in the Plains region. It is hardy as far north as Massachusetts in the east and Iowa and southern Illinois in the Central West. It does not form a large tree. The usual rate of growth under good conditions is one-quarter to one-third of an inch a year in diameter. Height growth is slow after the first few years. It forms a branchy tree and needs severe pruning to attain good form. The heartwood is yellow in color, heavy, tough, hard and strong. It is of value for cabinet purposes and for the manufacture of carriages, machinery and tool handles. The wood is extremely durable in contact with the soil and the fuel value is high. It has been planted extensively for hedges and windbreaks. The green "oranges" or fruit can be collected and the seed separated by soaking in water until the pulp ferments. The seed should be dried and stored in a cool, dry place over winter. The seed germinates readily when planted. In starting plantings the trees should be set closely together to overcome the branchy nature of the tree. They are often planted 2 feet apart in rows 8 feet apart. It is often mixed with cottonwood, black walnut and hardy catalpa on rich, moist soils and with black locust, honey locust, white elm or green ash on dry upland soils. The tree is usually free from serious attacks either by insects or fungus diseases.

Csage orange grown as a hedge and for the production of fence posts.

Red Oak (*Quercus rubra*).—This is one of the largest trees in the forests of the Northern States. Trees 150 feet in height and 5 feet in diameter have been found. The trees usually average 70 to 90 feet in height and 2 to 4 feet in diameter. It occurs naturally scattered over the eastern part of the country, running as far south as Georgia on the Appalachian Mountains. It will grow well on all soils of medium quality. It does best on bottomlands and lower slopes where it gets some moisture in the soil. It will thrive on heavy clay soils. The wood is heavy, hard, coarse grained and strong. It is in demand for lumber in many regions. The tree makes fast growth for an oak and it would be one of the best to grow for fence posts and ties if the wood were more durable. It absorbs creosote readily, however, and when treated will make one of the most durable of posts or ties. The trees reproduce both from sprouts and from acorns. The acorns should be planted in the field where the tree is to grow. Three acorns should be planted in spots about 5 to 6 feet apart. The acorns can be treated with red lead to prevent destruction by rodents. This can be done by dampening the acorns and shaking in a bag with a small amount of red lead. The acorns should be planted about 1½ to 2 inches deep. Under good conditions as on bottomlands and lower slopes, red oak at 20 years of age will reach a height of 25 to 30 feet and a diameter breast height of from 3 to 5 inches, and at 40 years a height of 40 to 50 feet and 8 to 10 inches in diameter. Being a fairly tolerant tree it should be of value for underplanting old stands of timber and for mixing with European larch, white pine and other trees.

Russian Mulberry (*Morus alba tartarica*).—The Russian mulberry is a foreign tree that has proved successful in the semi-arid regions, owing to its ability to endure almost any amount of drought. It will grow on either sandy or clay soils. It does not grow to large size, but under favorable conditions produces very rapid growth. A growth of an inch in diameter and 2½ feet in height has been attained, though usually of much slower growth. It is suited to the climate of southern Nebraska, Kansas, Oklahoma and Indian Territory. It cannot endure severe winters. It forms naturally a very low bushy tree, and requires constant pruning to produce a good trunk. The wood is heavy, hard, elastic, coarse grained, and moderately strong. It splits easily and has high fuel value, and makes a durable fence post. It is often used for windbreaks and hedges. Its fruit furnishes food for birds. It reproduces by seeds, by sprouts and by cuttings. The seed can be separated by crushing and washing the berries. The seed can be kept dry over winter, but it is better to stratify it in moist sand. It should be planted 4 by 4 feet for lumber and 2 feet apart for windbreaks.

Shagbark Hickory (*Hicoria ovata*).—Shagbark hickory grows throughout the eastern portion of the country, scattered throughout stands of other trees. It never forms pure stands, though often occurring pure in clumps or groups. It grows to large size, often reaching 130 to 140 feet in height and 20 to 30 inches in diameter. It prefers a fresh, fertile soil. It will grow on fairly dry soils, but for best growth it demands some moisture. The wood of hickory is heavy, hard, strong and tough. The wood is straight grained and splits readily. It is used where toughness and strength is necessary, as in vehicles, handles, axes, automobile and wagon rims, shafts and spokes. It has the highest fuel value of any North American wood. The growth is rapid on good soil after the first five years, a foot a year for 60 years often being made. In diameter the average increase is an inch in 7 years. It has long life, often living for 250 to 300 years. It is a fairly prolific seed-bearer and is easily propagated by planting the nuts. It is also a good sprouter in youth. The nuts should be stratified in moist sand over winter and planted where the trees are to grow, because the tree produces a deep taproot. Hickory is often managed by the coppice system on short rotations for the production of sprouts used for making handles and spokes.

Pignut Hickory (*Hicoria glabra*).—Pignut hickory is distributed throughout the eastern part of the country. It usually occurs mixed in with other species. It often reaches a height of 130 to 140 feet and 3 to 4 feet in diameter. It forms a very strong taproot like the other hickories. It grows both on clay and on sandy soil. It prefers a moist, fairly fertile soil. It succeeds better on drier situations than the other hickories. It grows at about the same rate as shagbark hickory, but continues rapid growth for a much longer period. It produces the strongest and toughest wood of all the hickories. It produces seed in abundance, and since they are rarely edible, natural reproduction is good. In addition it is the most vigorous sprouter of all the hickories. It is propagated and managed the same as the shagbark hickory. Owing to its ability to endure fairly dry conditions, it is a valuable tree for planting on many situations.

Sugar Maple (*Acer saccharum*).—This tree grows naturally in the Eastern United States. It often reaches a height of 120 feet and a diameter of 4 to 6 feet. It prefers a fresh, well-drained soil. It grows well on any rich soil. It will not thrive on poor, dry ground. The wood is heavy, strong, dense and hard, but not durable in contact with the soil. It absorbs creosote readily, however, and when treated posts will last 15 to 20 years. The wood is in demand for many purposes. The tree is grown extensively for the production of maple sugar. Since the amount of sugar produced depends on the extent of the leaf surface, the trees for sugar production should be spaced wider apart than for the production of wood. Six feet apart is the usual spacing for wood production. The tree grows well in pure stands and can also be mixed with other trees, such as white pine, red pine, European larch, red oak, yellow poplar, etc. The trees are easily propagated from seed. The seed should be gathered from the trees and stored over winter by stratifying in moist sand. The tree is rather slow-growing, but of long life. Under favorable conditions of soil and situation at 30 years of age trees should reach 35 to 40 feet in height and 6 to 8 inches in diameter breast height.

Silver Maple (*Acer saccharinum*).—Silver maple is widely distributed over the eastern portion of the country. It thrives best on river bottoms. It is a moisture-loving tree and fails when planted on dry uplands, but it wants a fairly well drained soil. The tree grows rapidly, attains large size, but is short-lived. It often grows a half inch in diameter a year. The wood is neither strong nor durable. It is easily worked and is sometimes used for flooring and furniture. It is used largely for fuel. It is often planted for shelter belts because of this fact, and also because of its fast growth. Because of its brittle wood it is often broken by storms. It reproduces by sprouts and seeds. The seed ripens in May or June and should be planted immediately in drills in the nursery, the seed being sown thick, for it is rather infertile. The seedlings will grow 1 to 3 feet high the first season, and should be set out the next spring. They should be spaced about 6 by 8 feet apart. On dry situations it can be mixed with osage orange or box elder. For moist situations it can be mixed with white willow, cottonwood, white or black ash, black birch or walnut. It is a hardy tree and comparatively free from insect pests or fungus attacks.

White Ash (*Fraxinus Americana*).—White ash is native to the Eastern United States. It is a tall, slender tree, reaching a height of 100 feet and a diameter of 2 to 3 feet. It prefers and makes its best growth on a rich, moist soil, such as bottomland or lower slopes. It will grow, however, under less favorable conditions. It will not grow in stagnant water. The wood is of high economic value. It is hard, strong, heavy, tough and elastic. It is used in the manufacture of farm tools and for baskets, handles, interior finish

Chestnut trees killed by the chestnut bark disease.

and furniture. It is fairly durable in contact with the soil. The rate of growth is rapid, but varies with the conditions of moisture and situation. Seedlings can be started from seed planted in the farm nursery. The seed should be gathered in the fall and hung in a bag in the barn over winter. The seed should be planted as early in the spring as possible, sown thickly in drills half an inch deep. The seedlings will be ready for planting the following spring. It can be planted pure or mixed with other species such as European larch, black cherry, black walnut, pine, etc. The trees should be spaced 6 by 6 feet on favorable sites, but closer together on poorer sites. At 40 years of age on good soil white ash will reach a diameter growth of 8 to 10 inches and a height growth of 50 to 60 feet. It can be planted for fence posts. When creosoted the posts will last many years.

White Willow (*Salix alba*).—White willow was introduced from the old world early in the settlement of the country. It thrives throughout the northern and eastern parts of the country and in most places has run wild. It reaches a height of 80 feet and a diameter of 4 to 6 feet. It prefers a moist or wet, rich, alluvial, sandy loam, but will thrive well on the high, dry prairies of North Dakota and Minnesota. It is a rapid-growing tree. It reproduces naturally from seed, but the usual method of propagation is by cuttings from one to two-year-old wood, made 8 to 10 inches long. These are planted firmly in the ground so that only two buds appear above the surface, and spaced 3 feet by 6 feet or 2 feet by 8 feet apart. It is sometimes mixed with cottonwood. Weeds must be kept out until the ground is well shaded by the crowns. The wood is very soft, flexible and fairly strong. It is used for cricket or baseball bats, in turnery and cooperage. The charcoal is said to make the finest grades of gunpowder. It is sometimes used for fence posts, although it has but little durability. When planted in dense stands it yields straight, light poles for many purposes. It is also planted for windbreaks.

Because of its extensive root system it is well adapted for holding the soil along streams and to stop erosion. Many different kinds of insects live on the tree and often cause considerable damage, and the wood is sometimes injured by heart rot and other fungus diseases.

Yellow Poplar (*Liriodendron tulipifera*).—This tree is a native of the Eastern United States and is hardy as far north as Southern New England. It is comparatively free from insects and disease. It has fast growth and high quality of wood. The tree is known as yellow poplar, tulip poplar, and when young as white wood. No other tree in the woodlot produces such a clear, straight, cylindrical trunk. The tree grows to be 125 feet in height and 3 to 6 feet in diameter and is sometimes much larger.

Yellow poplar is exacting as to soil and moisture requirements for good growth. It demands deep, fertile, well-drained soil with a constant and even supply of moisture. It does not thrive on shallow, dry soils or on ridges or in standing water.

The wood is light, soft, tough, easily worked and of fine texture, but not strong. The brownish yellow heartwood is fairly durable. The wood is usually cut for lumber, and is valuable for many purposes, such as construction, interior finish, woodenware, crates, etc.

Yellow poplar can be raised from seed. The cones of seeds should be collected from the trees and stored over winter by stratifying in moist sand. The seed should be sown thickly in rows, for the seed is very infertile. The seedlings should be planted after one year growth. In planting it is best to mix some other species, as white or red pine, European spruce and European larch, because it does not do well when planted alone.

Yellow poplar is one of the fastest-growing hardwoods on good soil. At 40 years of age on good soil the diameter breast height will be from 10 to 12 inches and the height from 60 to 70 feet.

CHAPTER IX

NATURAL METHODS OF STARTING AND REPRODUCING A WOODLOT

WOODLOTS are reproduced naturally, as has been said, by sprouts from the stumps of trees when cut and from seed sown from trees. Most of our woodlots have been produced naturally in one or the other of these ways, except in the treeless regions. Natural methods of starting woodlots will be the ones relied on by farmers where reproduction can be secured, because they are much cheaper than artificial methods and the new stand is started without any particular trouble and often without expense or labor. In a region where trees naturally grow there should be little difficulty in starting a woodlot naturally from seed or from sprouts.

SECURING A NEW STAND OF TREES FROM SPROUTS AFTER THE OLD TREES HAVE ALL BEEN CUT, CALLED THE SPROUT OR COPPICE METHOD

Sprout or coppice growth is one of the most common methods and one of the surest methods of naturally reproducing a woodlot. The trees are all cut and the new stand starts from sprouts from the stumps. This method is applicable to broadleaf trees that sprout. Conifers or evergreen trees are reproduced naturally only from seed. There are but few of them that produce sprouts that grow to tree size such as the redwood in the West and the shortleaf, loblolly and pitch pines in the East. Some broadleaf trees will also start from sprouts from the roots called root suckers, especially when the roots have been injured or bruised. Poplar, beech, wild plum and black locust are examples of trees that reproduce themselves by root suckers as well as by sprouts.

Trees differ in their ability to produce thrifty sprouts. Chestnut, black locust, willow, sugar maple, cottonwood, catalpa and red oak are among the best sprouting trees. Hickory, basswood, soft maple and ash also sprout well. Old trees beyond 60 years usually lose their power of sprouting, so a woodlot managed by this system must be cut early in life. The trees should be cut in winter or early spring before the new growth has started. If cut in summer sprouts will be produced, but they will be killed by frost during the winter. The stumps should be cut low so that the sprouts will not easily be broken or blown off. The stumps should be cut smoothly and in a slanting direction, so that they will shed water and keep from decaying as long as possible. The bark should not be injured or loosened from the stumps or the sprouting ability will be lost.

Sprout trees grow faster in early life, but rarely reach as large size as do trees that have started from seed. A sprout woodlot is used to grow poles, railroad ties, fence posts and cordwood. Sawlogs are usually grown from seed trees. After three or four generations of sprouts have been produced from the same stump, the trees produced will be less vigorous and have slower growth and shorter life. For this reason better trees will be produced from the stumps of seedling trees than from the stumps of sprout trees. So young trees that start from seed in sprout stands should be encouraged to grow to renew the trees and to keep the woodlot thickly stocked.

Not all the sprouts that start from a stump should be allowed to grow. Several of the best sprouts should be selected and the rest either cut or broken off the stump, or the top of each sprout broken to prevent further height growth. Later the number of sprouts should be further reduced to not more than two or three to each stump. Seedlings of trees are often planted in sprout stands to fill in between the stumps, which

The stump of a large tree producing but a few sprouts.

White pine seedlings started naturally from seed.

often stand far apart, so as to secure good density of trees. When a woodlot is cut clean a few of the more vigorous trees are often left to grow to large size while the new woodlot is being established by sprouts. These trees are usually of seedling origin and are allowed to grow until the new woodlot is to be cut. In this way sawlogs can be produced as well as smaller material. This is known as compound coppice.

A combination is sometimes made of sprout and seedling reproduction called polewood coppice, that is applicable to many woodlots. It is often used where there is not a good market for small material. The trees are allowed to grow to large size for making poles, piles, ties or lumber. Since these trees when cut will have passed the period of best sprouting capacity, a new stand would be difficult to obtain from sprouts. If such a stand is cut clean a very open stand will result. To secure a new stand the trees can be removed in two cuttings. A heavy thinning is made so as to open up the trees left standing to the light. Seed production will be stimulated and when a seed crop has been secured and seedlings result the remaining trees are cut. The new stand will be composed of seedling trees and what sprout trees have started from the stumps. Sometimes the woodlot is cut clean and seedlings of trees are planted among the stumps with the same result.

STARTING OR REPRODUCING A WOODLOT BY SEED SOWN FROM NEIGHBORING TREES, THE CLEAR-CUTTING METHOD

Sometimes a woodlot is entirely cut off and no provision is made for securing a new woodlot, but the trees standing about the area are relied on to sow seed for starting new trees. When coniferous trees are cut, like the pines and spruces, or when broadleaf trees cut, are old and beyond the age for producing vigorous sprouts, or where trees are cut in summer and no vigorous trees result from the sprouts put forth, a new woodlot must be started naturally from seed. Some tree seed is blown long distances. Light seeded trees on the edge of the area cut over will soon seed up the ground with new trees, provided the area cut is not too large and provided the soil is in good condition. A soil packed hard by the grazing of many cattle or a soil that has been repeatedly burned over, will not be in a good condition to germinate seed. Cut-over areas are often seen growing up to weeds and grass instead of trees, even when seed trees are growing along the edge of the tract, because of the poor condition of the soil. A field that has been abandoned will soon grow up to forest trees because of the good condition of the seed bed. Not only will light seeded trees seed up cut-over areas, but heavy seeds like walnut, hickory, oak and chestnut will be carried in by birds and squirrels. The squirrels gather large quantities of such seed, and besides storing the seed in their hoards in hollow trees and under logs, they will carry large quantities out into the open and bury them. A large part of this seed is never recovered and germinates and grows into trees. So an area cut over will soon be covered not only with trees having light seeds, but the trees having heavy seeds will gradually be introduced. Finally, all the trees that grow naturally in the region will be found growing on the cut-over area. If the area cut over is too large, not all of it will be seeded up satisfactorily with trees. Even where the area is small and there are many trees surrounding it to sow the seed, the reproduction may be too scattering in some places and too thick in others. Seedlings from the tree nursery should be planted wherever failures occur.

Sometimes the woodlot is cut off a strip at a time, the strips being not more than once or twice as wide as the trees are high. When the strip cut over is seeded with seedlings from the neighboring standing timber a new strip is cut, and so on across the woodlot. The first strip cut should run at right angles to the prevailing winds and on the side of the woodlot away from these winds, so that the seed from the trees left standing will be scattered over the cut-over strip. This is known as the strip method of reproducing a woodlot. Where single trees or groups of trees are cut from the woodlot, the area cut is soon seeded from the surrounding trees, provided the soil conditions are favorable.

STARTING A WOODLOT BY LEAVING TREES IN THE CUTTING TO SEED UP THE AREA, THE SEED-TREE METHOD

Often when a woodlot containing trees having light seeds is cut clean, several large seed trees per acre are left to seed up the area naturally to new trees. The number left will depend on the kind of trees and on the distance to which the seed is scattered. Usually two or three vigorous and wind-firm trees are sufficient if left well scattered over the area. When the trees are not naturally wind-firm, groups of trees are often left so that the trees will protect each other from the wind. This is known as the group seed-tree method. The seed-tree method is applicable to trees that are wind-firm and that have light, easily blown seeds that will germinate on clearings, like pines, spruces, ash, tulip, poplar and maple. Heavy seeds like walnut, oak, beech and chestnut do not scatter far, and this method cannot be used satisfactorily with such trees.

The success of this as well as of other methods of natural seeding will depend in large measure on the condition of the seed bed. If the seed is sown from the trees soon after logging and when the ground is torn up and exposed, natural reproduction will usually be successful. If grass and tall weeds take possession of the area, reproduction will be hard to establish. Hogs are sometimes driven into a woodlot that is to be reproduced naturally by seed to root up the ground and expose the mineral soil.

STARTING A NEW WOODLOT BY REMOVING THE OLD TREES IN TWO OR MORE CUTTINGS, THE NEW STAND STARTING NATURALLY FROM SEED SOWN BY THE TREES LEFT AND UNDER THEIR SHELTER, KNOWN AS THE SHELTER-WOOD SYSTEM.

According to this method about half the old trees are removed at the first cutting. The trees to be cut are taken here and there throughout the woodlot so as to break the canopy uniformly and allow the sunlight to fall on the forest floor. The trees left will soon start to grow vigorously after the thinning, because the increased sunlight will enable the trees to manufacture more food. The sunlight falling on the floor will help to decay the leaves and twigs and furnish additional food for the trees. The result will be a heavy crop of seed. The seed falling on the ground under the shade and shelter of the old trees will germinate, and soon the floor will be covered with young seedling trees. The new stand may be secured at once, or it may take several years to bring about, depending on the seed crop. A woodlot does not produce heavy crops of seed every year.

After the young seedlings are well started the rest of the old trees are removed. Many of the seedlings will, of course, be destroyed by the cutting of the old seed trees, but enough should escape to produce a well-stocked stand. Sometimes when the trees are very thick in the old stand and so are not wind-firm, and the removal of trees around them would be apt to cause them to be blown over, several light cuttings are made instead of one heavy cutting, removing but few trees at a time. In this way the trees left will gradually thicken and strengthen their root systems and become wind-firm, and be able to withstand storms without falling before they have sown their seed.

In all these different methods it will often be found that seedlings of trees have already begun to grow in the woodlot. This is called advance reproduction and should not be injured.

Trees left scattered over the area in cutting to sow the seed for the next crop. Norway pine in Minnesota.

Medium-sized trees, however, should not be left standing, for they will grow a wide-spreading crown and occupy the space several trees should occupy. Any advance reproduction present should be saved from damage and allowed to develop with the new seedlings obtained.

CHAPTER X

CARING FOR THE GROWING WOODLOT

WHETHER a woodlot is started artificially or naturally, it will need care and attention throughout its life, if it is desired to grow the greatest amount of valuable wood material in the shortest time possible. Nature must be assisted in her efforts by giving to each tree the conditions necessary for its best growth and development. When a growing woodlot is left entirely to nature the result will be what nature would be expected to produce, a stand of slowly growing, irregular and usually defective trees. A stand of growing timber is a crop the same as any farm crop, and will need oversight and attention. If this is given the results are surprising, not only in the character of the trees, but in the much shorter time it takes to grow them.

Cultivation in a Forest Plantation.—Cultivation of the ground around the trees after planting is advisable. It is necessary in the treeless regions of the Prairies and Great Plains and wherever the conditions for tree growth are not favorable. Transplanting is a severe shock to trees and even under the best of conditions many often die. With proper care and cultivation during the first few years after planting a much higher percentage will live. Cultivation of the soil will conserve the soil moisture, will prevent the growth of grass and weeds, will help the trees to become established early, lessen the mortality among the planted stock and shorten the rotation. All these points are of particular importance where trees are being grown for fence posts or other material for sale. In the Eastern Region under more favorable conditions of

A young stand of hardwoods in which a cleaning should be made to remove weed trees and to determine the trees that should grow.

trees and seldom results in a revenue to the owner, owing to the small size of the material removed. The work of cleaning can be carried on in the winter time. It will repay the owner, in that the trees will grow faster and the poorer kinds of trees will be removed, assuring a stand of desirable trees in the future.

In making a cleaning only sufficient material should be removed to accomplish the purpose desired. Sometimes a tree overtopping a better tree can be broken off or bent to the ground so that it will no longer injure the tree beneath. Sometimes the cutting of a single branch will release to the light the tree being shaded. Only material should be removed that is necessary, so that the cleaning should be made rapidly and at little expense.

Pruning.—Pruning is the cutting of lower dead or living branches from trees. The objects of pruning are to improve the form of trees, to concentrate the growth in a single leader, and to secure trunks clear of limbs, so that lumber free from knots will be grown. Pruning would not be a practical operation over a large forest, but in a small farm woodlot or forest planting it can often be done with great advantage to the trees and profit to the owner. Pruning takes place naturally in the woodlot where trees stand closely together. Young saplings are often seen with long, clear boles, free of branches for many feet. With intolerant trees this natural pruning takes place rapidly. With tolerant trees the lower branches will remain alive many years and will remain on the tree many years after they have died.

In pruning limbs they should be cut off smoothly and as close to the trunk as possible without tearing the bark. With large limbs it is best to make an undercut first, so that the limb in falling will not tear the bark of the tree. Limbs that are dead and the wood brittle can be knocked off. The trunks of the most vigorous trees in the woodlot should be pruned of branches for from 16 to 20 feet, so that at least one good moisture and growth, cultivation of the trees helps them to become established and hastens their growth during the first few years. Field crops can be grown between the rows of trees, so that the entire cost of cultivation need not be borne by the tree crop. The number of cultivations a year and the number of years the plantation should be cultivated will depend on the rapidity of growth of the trees, the spacing used, the character of the soil and situation and the climate. Cultivation of forest trees, as with fruit trees, should not be continued late in the season for fear of the trees being winter-killed. With late cultivation the wood of trees continues to grow and remains soft and full of moisture. Freezing will kill back such tender shoots. Cultivation should not be continued beyond midsummer. When the trees have been started from seed sown by the trees they will seldom need cultivation. The young trees starting from seed send out their roots in all directions and become well established at once, so that usually they can compete with the grass and weeds that grow with them unless this material becomes too high.

Cleanings in a Growing Woodlot.—A cleaning is a thinning out of trees in a very young stand. Its object is to improve the stand and to determine early in the life of the stand the trees that are to grow. It is usually made before the trees are 10 years old. Often the better kinds of trees will be outgrown by inferior ones or be crowded by trees of poor form, and unless these are removed they will interfere with or cause the death of the better trees. Sprouts, because they grow much faster than seedlings for the first few years, will often overtop the seedlings and destroy them by their shade. When the seedlings are wanted the sprouts interfering with them should be removed. Often in natural reproduction the young seedlings will be bunched very closely together in a dense growth. The removal of some of the seedlings will greatly benefit the growth of those left.

A cleaning is made for the benefit of the young growing

pole or sawlog can be secured. In some cases only dead branches are removed, but often lower live branches can be cut to good advantage. Care must be taken not to reduce the crown materially. In small fast-growing trees wounds will grow over quickly before the wood begins to decay. If the wounds are larger than a square inch or two in size they should be painted with some good paint until the wounds have been covered with new bark. Coniferous trees usually cover over such wounds with pitch.

Thinnings Made to Increase the Rapidity of Growth of Trees.—If as the trees in a woodlot grew the crowns did not expand, the young woodlot could be left entirely to nature just as a farmer leaves his wheat to grow by itself. He sows a certain number of seeds to the square foot and he has learned by experience that when planted so far apart individual plants will not interfere with each other. But trees do not grow as do wheat and other grains which grow for a year only. The older a tree becomes and the higher it grows the wider it attempts to spread its crown. A tree will grow rapidly to a certain height only with a restricted crown. Unless room is then given for the crown to spread sidewise the subsequent height growth will be very slow. A thinning is made in a young growing stand so that the density of the crowns will be reduced and the trees left will be able to expand their crowns, and so carry on rapid height growth. It is made for the same reason that a farmer thins out his beets or other vegetable crop. If he allowed all to grow that came up there would be few beets produced. He thins out the plants so that the few that are left evenly spaced will have room to expand and develop.

The competition between the trees in a young woodlot becomes very great when it is near the end of the large sapling stage. So severe may be the struggle that the trees will often cease to grow, all the energy of growth being used up in contending with neighboring trees. Often a stand of young trees which are grown too closely together will show scarcely any growth for many years. This is because the crowns being crowded are unable to manufacture sufficient food and the roots are unable to absorb sufficient food and moisture for rapid growth. A thinning is made to lessen this competition or struggle between the trees. The trees to remove are those with the smallest crowns or those that have fallen behind in the struggle for existence. Not all trees have equal vigor or grow at the same rate. Some will in time forge ahead of their fellows, while others will lag behind. It is those that are falling behind, the suppressed trees or badly crowded trees, that should be removed in making thinnings, leaving the more vigorous trees or dominant trees to grow. Where all the trees seem to be growing equally well in a dense stand some of them must be removed to give the crowns of those left more room to develop. The tendency will be to remove too many trees. Only enough trees should be taken out to liberate a little the crowns of the trees left. In general the openings should not be so large but that they will close again in from 3 to 5 years by the growth of the crowns remaining. The struggle for existence must not be stopped entirely, for it is this struggle that produces the long, slender boles of trees in the woodlot. The size of the opening that can be made will depend largely upon the rapidity of the growth of the trees. Rapidly growing trees such as cottonwood or silver maple should have their canopy opened up to a much greater extent than slower-growing trees such as ash, oak and walnut. In young stands the trees to be removed can be selected from those that should remain by shaking the trees, so as the better to see the crowns. By thinning out the trees lightly in this way there can be accomplished in a short time what it would take the trees themselves years to bring about, that is, the death of the trees that gradually fall behind or the natural thinning of the stand.

Subsequent Thinnings.—The trees left after the first thin-

A dense stand of timber in which a thinning has never been made. The crowns of the trees are too narrow for the height, resulting in slow growth. A thinning should be made to permit the crowns of the trees left to expand.

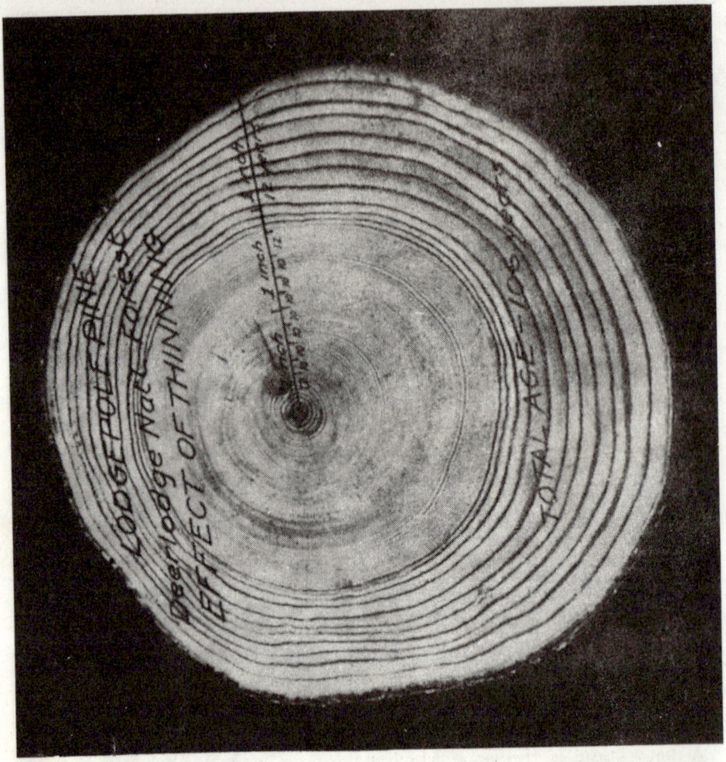

The result of a thinning in a dense stand of lodge pole pine shown in the increased width of the annual rings.

ning will start to grow faster as soon as the crowns are liberated. It will not be many years, owing to the spreading of the crowns of these trees sidewise, before their crowns will come together again and the crown canopy will be intact. The trees will have grown considerably in height when this takes place and this canopy will be many feet higher above the forest floor than it formerly was. As soon as the crowns touch the struggle starts over again between the trees. The time when this new struggle begins after the first thinning will vary with the kind of tree and the severity of the first thinning. In general in about ten years after the first thinning another struggle will have started. This older stand will live over again the life of the younger stand. The more severe the fight the slower will the trees grow. Soon some trees will begin to forge ahead and others to drop behind. The tree crowns again become separated into dominant, intermediate and suppressed trees. If the struggle is allowed to go on the trees will practically cease to grow, the energy being used up in the struggle for existence. It is time another thinning was made. The trees that have fallen behind and whose crowns have been overtopped should be removed to give the better-formed and taller trees more room to expand. Care must be taken, as before, not to remove too many trees so as to break the crown canopy too much. The struggle must continue for the production of clean boles. Owing to the large size of the trees when this second thinning is made the trees cut out can be worked up into cordwood. A cleaning and the first thinning rarely pay for themselves, as the material obtained is small. They are made for the benefit of the growing timber. The second and subsequent thinnings should pay for the time and labor needed to make them in the cordwood obtained.

Throughout the life of the growing stand every ten or fifteen years similar thinnings should be made. By relieving the stand of the necessity of killing out the trees that fall behind and by giving each tree the room it needs for its best and fastest growth, much time can be gained, and trees will grow to cordwood or standard size in a very much shorter time than if they had been left to themselves. By making thinnings in a dense stand of young trees we are helping nature to do in a shorter time what she would ultimately bring about unaided in a much longer time. By judicious thinnings throughout the life of a stand of growing timber often as much as twenty years can be cut off from the time necessary to bring the trees through to maturity. When the trees are finally mature and ready for the harvest but a few will be left of the many that started. Those that survive and form the final stand will be fine specimens of trees, the survivors in the struggle for existence.

Thinnings to Improve the Character of the Woodlot.—Woodlots that have not received proper care will often be found full of defective, ill-shapen and weed trees. A thinning made to remove this material from the woodlot to benefit the trees remaining is called an improvement thinning.

In removing this material from the woodlot care must be taken not to remove too many trees at one time. As far as possible the openings made in the canopy should be small, so that the neighboring crowns will close over them in a few years. The improvement should be a gradual one. The wood needed for use on the farm should be selected from the poor material year after year until it is all removed. A small amount removed each year for several years will accomplish the same result as many trees removed in one year, and their removal in this way will be an advantage to the woodlot. If too many trees are removed at one time the woodlot will deteriorate because of the exposure of the soil to the sun and other elements. One of the first principles of growing forest trees successfully, as we have learned, is in keeping the forest floor dark, so that grass and weeds which dry and exhaust the soil will not grow. If the canopy is broken too much by the removal of too many trees in a growing woodlot those

The ordinary form of farm woodlot in need of an improvement thinning.

Young coppice growth of white and black oak. The stumps standing far apart show the necessity of planting small trees to secure a fully stocked stand.

that remain will be apt to develop large limbs and the trees become of poor form.

The trees to select for cutting in the woodlot in order to improve its character are those that the woodlot can well do without.

Large Limby Trees.—In nearly every woodlot large trees with long limbs can be found that spread over a large space. They are sometimes called wolf trees. They are usually of slow growth, having passed the period of greatest vigor. Such trees ought to be removed, for they occupy the space many vigorously growing trees might occupy. Usually they are allowed to grow from year to year because of the difficulty of cutting them out and the labor necessary to work them up into cordwood. The removal of such trees is of great advantage to a woodlot. It enables several trees to start and grow in the place of the tree removed.

Defective Trees.—Often trees can be found whose crowns have been badly broken by winds or snow or other causes. Owing to the reduced size of the crowns these trees will have but a slow growth and there is danger of decay entering the trees through the broken parts. They should be removed to make way for others.

Misshapen Trees.—The trees that should be encouraged to grow in the woodlot are those with straight full boles. Such trees should be kept to grow to large size. Trees with crooked or gnarled trunks, or those that fork close to the ground, or that have been struck by lightning or are otherwise misshapen, should be cut before those of better form.

Stagheaded or Spiked-topped Trees.—Often trees will be found with dead tops, due to one cause or another. These have usually passed their prime. The dead tops show that they are no longer growing vigorously. They should be removed, to allow younger and faster trees to take their places.

Diseased Trees and Insect-infested Trees.—When trees become weakened from any cause they become the prey of insects and diseases. Such trees rarely make satisfactory growth and they are also a menace to the trees around them, because of the danger of the disease spreading. Trees that have begun to decay or that are badly attacked by insects should be immediately removed.

Weed Trees.—Every woodlot contains many species of trees that are of little value or of no use on the farm. These are called weed trees, such as dogwood, ironwood, blue beech, bird cherry or any kind of tree that does not serve some useful purpose. These trees occupy space in which more useful kinds of trees might grow.

Dead Trees.—Dead trees and limbs and fallen trees should be utilized for cordwood. Such material often serves as the breeding place for insects that may attack living trees. Where such material is utilized before it begins to decay it will make satisfactory fuel wood.

TREES THAT SHOULD NOT BE CUT, BUT SHOULD BE ALLOWED TO REMAIN IN THE WOODLOT

No tree should be removed unless there is some good reason for cutting it or unless it is needed for a special purpose, and no tree of less value will serve that purpose.

In general, young, thriftily growing trees should not be cut until mature. They are producing a high rate of interest in the amount of wood laid on each year, and are increasing rapidly in value.

In the ordinary farm woodlot of irregular form no tree should be removed unless there are smaller trees beneath it to grow up and occupy its place or unless some provision is made for starting new trees. Many woodlots have large holes in the canopy where trees were removed many years before. Grass forming under the openings prevents young trees from starting. Such openings should be planted with trees.

A tree should not be removed if its shade is needed to protect the forest floor or if its removal would expose large areas to the action of the sun and rain. Trees needed for sowing seed should not be removed.

CHAPTER XI

CARING FOR THE WOODLOT STARTED FROM SPROUTS AND THE ORDINARY FARM WOODLOT

Caring for the Sprout Woodlot.—The method of caring for a growing woodlot started from sprouts does not differ materially from the method of caring for a forest planting. The same operations of cleaning, pruning, planting up and thinning the stand as it develops are carried out.

Seedling trees of valuable kinds that have started among the stumps should be favored in making a cleaning. As we have already noted, sprouts from trees that have started from seed make more vigorous trees than sprouts from trees that started as sprouts, and that in time through several generations sprout trees may lose their sprouting capacity. Where sprouts, even of the better kinds of trees, threaten to overtop seedling trees of valuable species, the sprouts should be removed or their growth otherwise impeded, as by breaking off the leaders.

Where the stumps of the trees cut, stand far apart and there are no seedling trees coming up to fill in the spaces between, trees should be planted. Planting of trees or the making of seed spots is often a necessary operation in a sprout stand. If such areas are not planted with valuable kinds of trees inferior species will finally take possession of the soil or the woodlot will not be well stocked with trees. In planting in a cut-over woodlot large stock must be used, so that the roots will be well below the upper layers of the soil, which, because of the large amount of humus, will dry out quickly. Trees at least three or four years old should be used.

A stand of timber that has been properly thinned. The products of thinning pay for themselves in the cordwood or other material obtained.

A well managed stand of timber. Small trees left to grow to larger size and to sow seeds for a future crop. Brush piled so that it can be burned to lessen the fire risk. A continuous yield assured.

In time the crowns of the different sprouts and of the trees that have started from seed or that have been planted will come together and form a crown cover or canopy. From then on the life history of the sprout stand will be similar to that of a forest plantation or even aged stand, and it should receive the same treatment. As soon as the competition between the trees becomes strong, a thinning must be made to release the crowns of the better and more vigorous trees, and similar thinnings should be made throughout the life of the sprout stand whenever needed.

CARING FOR A WOODLOT THAT HAS STARTED NATURALLY FROM SEED SOWN BY TREES

When the woodlot has started from seed sown by neighboring trees or from trees left for the purpose of sowing seed, the resultant stand will be apt to be more or less irregular. Near the seed trees the ground may be well covered with young seedlings. If the seedlings stand 5 or 6 feet apart each way the reproduction will be satisfactory, for that is as closely together as trees would ordinarily be planted in starting a forest planting. Usually there will be many more. There will be many areas where the seedlings will be far apart or where no seedlings will have started. Such areas should be planted with seedlings from the farm nursery. It is essential that the woodlot be well stocked from the start if trees of good form and of value for use are expected. Openings in a woodlot usually persist. The ground becomes covered with a thick sod or a mat of weeds and bushes, that prevent subsequent seeding of trees.

A cleaning ordinarily must be made sooner in the life of a naturally produced woodlot than in one in which the trees are planted and evenly spaced from the start. This is because wherever reproduction has been good the trees will stand thickly together, often forming thickets through which it is almost impossible to pass. The competition between the trees in such dense stands begins early and should be stopped by thinning out the less likely trees. In many woodlots before the old trees are cut young seedling trees will get started beneath them and grow. Especially is this true where the trees do not stand closely together allowing considerable light to fall on the forest floor or where the trees are old and the crowns thin. When the old trees are removed this advance reproduction is present and will start to grow under the stimulation of increased sunlight. Where this advance reproduction is of desirable species and does not interfere too much with the new reproduction it should be encouraged. Often such advance reproduction is of poor species of trees, or because of its greater age and size it may interfere with or kill out many seedlings of more valuable kinds. In this case the advance reproduction should be cut out, liberating the seedlings beneath it. Such a cutting is known as a liberation cutting. Where advance reproduction is of value for shading or protecting the soil it should be allowed to remain.

When the young stand has grown up so that the crowns meet and the floor is shaded the life of the woodlot as a unit begins. Its subsequent treatment will be the same as for planted stands. Thinnings must be made throughout its life whenever the competition between the trees becomes too strong or whenever it is desired to improve the character of the stand.

CARING FOR THE ORDINARY FARM WOODLOT

The ordinary small woodlot differs in form from the even aged woodlots that start from planting seedlings or that grow from sprouts or from seed sown from trees soon after the woodlot has been cut. The ordinary farm woodlot is composed of trees of all sizes and ages mixed in together. It does not have a single crown cover as do even aged woodlots. The crowns of trees stand at all elevations, from those of one-year-old seedlings to those of poles or veterans. We have

The natural woodlot. The result of lack of care and attention. The woodlot filled with old, defective, broken and dead trees.

The ideal form of the small farm woodlot. Trees of all sizes and kinds mixed in together.

learned that such a woodlot has an irregular form. It originates usually from the method of cutting the trees. Where the woodlot is cut clean all trees being removed and the new growth starts at once, an even aged stand results. If instead of cutting all the trees at once the owner cuts out a few of the larger trees year after year, an irregular woodlot will result. When a large tree is removed there will be an opening formed in the canopy. Little seedlings under the opening will be stimulated to growth by the increased light and will at once start to fill in the vacant space in the canopy. In this way the woodlot will in time become filled with trees of all ages.

Because the owner in cutting the trees makes a selection of those to cut out, choosing certain ones from those surrounding them, this method of reproducing a woodlot is called the selection system and the irregular forest that results is called a selection forest. On large forests the owner may decide to cut out all trees down to a certain diameter, saving for future growth all the small trees below that diameter. This is also a form of the selection method.

The future of the irregular form of woodlot depends, of course, on the young seedlings that start to grow under the old trees or in openings that are caused by the removal of those trees. They are the children of the woodlot. If they are destroyed year after year by fire or grazing or if they are cut out thoughtlessly as the owner thinks to improve the woodlot, there will be no trees to grow up and take the place of the older trees when they are removed. In no form of woodlot is it more necessary to take good care of the young growth than in the irregular form. In all even aged woodlots after the crowns of the trees are well up above the heads of cattle, a little grazing may do no immediate harm to the trees, though it would be better for the trees if grazing were excluded. But where the maintaining of the forest capital and the future of the woodlot depends on the young seedlings and saplings that start each year in the woodlot, they must be protected and encouraged.

Where there are no younger trees or seedlings beneath the older trees and no seedlings start in the openings when the trees are removed after waiting a few years, seedlings of trees should be planted. Trees should be grown in the farm nursery for this purpose and they should be of larger size than required for planting in the open. They can be allowed to grow to large size in the nursery, so that when planted out they will soon grow to trees of value. Often young seedlings can be taken from some other portion of the woodlot and planted in the openings. Along the edges of woodlots or around openings there are often many trees growing that will soon begin to crowd each other. Some of these trees can be utilized for this purpose before they die. It is usually necessary to plant trees in the woodlot to keep it well stocked. By planting the trees or the seeds of trees is the only way to obtain trees of valuable kinds. There are many kinds of trees that are far better to grow in woodlots than those many times found growing.

All the various cultural operations described for other forms of woodlots can be carried out to good effect in the irregular form. Where valuable kinds of trees are planted to fill vacant spaces it may be necessary to cultivate the soil. A thick, heavy sod sometimes grows in such situations. This may deprive the trees set out of moisture and food before they get a good hold on the soil and so they will die. Cultivating the trees will make them more sure of withstanding the transplanting and will make them grow faster during the first few years so they will keep ahead of the weeds. Cleanings are often necessary where many trees grow up together. Where inferior trees begin to crowd or overtop more valuable kinds they should be removed or their tops broken to retard their growth. Often sprouts from the stumps of the trees cut where very numerous will interfere with the growth of

The wornout woodlot. The old trees should gradually be removed and utilized and the woodlot restocked by planting trees.

each other and will need to be thinned out. Or the sprouts that start may because of their faster growth begin to crowd more valuable seedlings and should be removed. Thinnings to increase the rate of growth of the older trees as well as the younger trees where they stand too thickly should be made, the trees that are falling behind being removed and the more vigorous trees given more room for their development. Thinnings to improve the character of the stand should be made every few years, the trees to cut for cordwood being chosen from those the woodlot can well do without. There is no owner of forest land who can practice forestry more intensively or to better advantage than the farmer on this kind of woodlot.

THE WORNOUT WOODLOT

Wornout woodlots are those in which the forest capital or the number of trees has been greatly reduced from the normal number that should be present. The trees stand far apart and no longer shade the floor. Grass, weeds, berry bushes and shrubs cover the ground. The trees are nearly all old and more or less defective. Such a condition results from removing too many trees year after year, from the constant removal of the fastest growing and more vigorous trees and leaving the poorer ones, or from grazing or fire that have killed the young trees and so have destroyed the future of the woodlot. Such a woodlot is of but little value for a farm. The trees being large will not supply the many kinds and sizes of material needed. It is often turned over to grazing.

A wornout woodlot needs to be restocked by planting. There is no other sure way of increasing the forest capital. The ground among the trees might be fertilized and cultivated and the trees thus stimulated to seed production and the soil put in good condition to receive the seed when a seed year is present. But it is difficult to renew a wornout woodlot by natural reproduction. The trees are too old to produce seeds abundantly. Should a seed year come, and it might be several years before seed would be produced, the seed might be infertile or be eaten by squirrels. The only practical and sure method is to gradually remove the old trees as they are needed for cordwood or other purposes and to plant trees of desirable species that have either been raised in the farm nursery or transplanted from some other woodlot. Seedlings can also be obtained from nurserymen. It is not necessary to plant the whole woodlot in one year. An acre or less planted each year will soon cover the woodlot with young trees.

The planting of the seed of trees in the woodlot is one of the best methods of keeping up the stock of trees. Seed can be gathered in the fall of the year and either planted at once by thrusting a stick in the ground and planting the seed in the hole made or the seed can be stratified over winter and planted in the spring. Many seeds can be gathered in the spring of the year after the snow has disappeared, that have been stratified naturally on the ground over winter. Where squirrels are not numerous, acorns of the many species of oak, walnuts, butternuts, hickory nuts, beech nuts and chestnuts can be gathered and planted in holes made in the ground. Smaller seeds like maple, ash, yellow poplar and locust can be gathered and planted in seedspots in the woodlot. Seed that has been stratified naturally over winter will start to grow at once as soon as the ground becomes warm. There is no more practical or effective way of starting trees growing in the woodlot.

THE GROWING OF BASKET WILLOWS

The growing of basket willows can be profitably carried on by farmers who have land that is subject to inundation or flooding by water. Such land is of little value for other purposes. The basket willow industry has developed rapidly in recent years and there has been a constant demand for willow rods.

The basket willow showing the character of rods produced by the Lemley willow.

The basket willow is a variety of the willow family that is especially adapted for this purpose. The common willows do not yield rods of high grade. Most of those in use have been imported from Europe. The willow is not a tree that demands water. It grows best on rich, moist, well drained bottomlands. It will thrive on land subject to inundation provided there is good drainage. It will not grow well under swampy conditions. If the land is flooded for long periods of time the willow will not live. The species commonly planted are the Welsh or purple willow, the Lemley, patent Lemley or Caspian willow, the American green or almond willow and the common white willow.

The basket willow is propagated by cuttings from one year-old shoots cut into sections 10 inches to a foot long. These are planted in rows in the willow bed or holt from 16 to 20 inches apart, the cuttings being spaced from 6 to 9 inches apart in the rows. The cuttings are either forced into the ground or set in holes made by a stick or sharp iron. They are planted upright with the buds up so that only an inch remains above the ground. Close planting retards the growth of weeds and results in greater yield per acre and in straighter and less branchy rods.

The land for a willow holt should be thoroughly cultivated as for a field crop and limed and fertilized with wood ashes. On land never flooded the soil should be deep and moist, but need not be necessarily rich. If land is subject to overflow it must be well drained. Such land does not need fertilizing. Ordinary soil must be fertilized every 4 or 5 years. Cultivation is necessary, for the willow is very intolerant of shade and is easily crowded out by weeds. Three or four hoeings must be given the first year and two the second. Later the willows will be thick enough to shade out the weeds.

The rods should be cut close to the stool. If cut below the ground the rods will be brittle. If cut too far out from the stool the rods will be apt to curve at the base. Rods to be sold with the bark on must be smooth, tough, flexible, branchless and cylindrical, and the color of the bark must remain a light brown. Usually the rods are peeled before being sold. These must have the same general characteristics and must remain white when peeled. In addition rods must have small pith and straight grain to bring the highest prices. Peeling is done either in the spring when the sap begins to flow and the bark slips easily, or the rods are steamed to loosen the bark. The rods for steam peeling are cut in the fall as soon as the leaves have fallen. Where sap peeling is used the rods are not cut until March.

After the rods are cut they are drafted or sorted into 4 or 5 height and quality classes. To do this they are placed upright in a barrel and all those of a certain length are placed together. The smallest rods usually bring the highest prices. The care taken in sorting the rods will often determine the price received. After being sorted the rods are tied in bundles by grades and stood in shallow pits of water about 2 inches deep. After standing for 2 to 3 weeks the ends of the rods become covered with little roots and the rods with tender green foliage. In this condition they are ready for peeling. In peeling, the rods are first drawn through an instrument called a brake, which bruises and loosens the bark. Strippers then pull off the loosened bark. The rods are then bleached in the sun and dried in the open air. They are then bundled tightly in bundles about a foot in diameter and stored in a dry, dark place.

A fair holt will average 4 tons of rods to the acre after the second year for 12 years. The first year rods are of little value. The second year the holt will yield about half a crop. The price varies with the kind, size and quality of willows from 4 to 6 cents a pound. The cost of growing a crop of willows will be about $50 an acre. An acre of willow will also yield about a ton of bark which should be utilized. It contains sufficient tannin to be put to commercial use. Abroad the bark is used for making doormats, for bedding cattle and for fodder. It sells for about $15 a ton.

MANAGING THE MAPLE SUGAR BUSH

In the northern states where the hard or sugar maple forms a large percentage of the trees in a woodlot, the making of maple sirup or sugar is an important spring industry. Many farm woodlots are managed especially for this product.

Sugar maple is a forest tree and to grow well in a woodlot requires forest conditions. The canopy must be kept intact so that the sunlight will not fall on the forest floor, and the ground must be covered with a layer of leaves and humus to the exclusion of grass and weeds. The amount of sap produced by a tree is proportional to the extent of the leaf surface. In the woodlot the best results will be obtained not by a few trees with very wide spreading crowns, but by the largest number of trees consistent with fully developed crowns. In the "sugar bush" the trees should stand farther apart than in the farm woodlot managed for the growing of wood products alone, in order that good crown development will be secured.

Maple, being a tolerant tree, would naturally grow in uneven aged stands containing trees of all sizes and ages. Such a stand would be managed by the selection system, the old trees being removed when mature and their places being taken by younger trees that have developed beneath them. Ordinarily stands of the sugar maple are more even aged owing to the trees having seeded in on a clearing or because of grazing that has killed off the younger trees until all the trees are about of the same size. Where such a stand is mature and the trees stand far apart, small maple trees should be planted among the trees and in the openings, and the mature trees gradually removed. In young, even aged stands where the trees stand closely together, they must be thinned out every few years lightly, so as to induce crown development. Certain thrifty trees with good crowns can be selected for the final stand and the trees around them when they begin to crowd can gradually be thinned out. Heavier thinnings can be made than in the woodlot where trees with long boles are desired. At 40 years of age there should be about 100 trees to the acre and the crowns should completely shade the forest floor. It is seldom that sugar maple forms pure stands. It is usually found in mixture with other trees. Where it is desired to develop a sugar bush these trees should gradually be removed by utilizing them for fuel and other purposes on the farm and their places in the stand filled by planting the sugar maple.

The sap season begins the middle of March and continues to the third week in April. The season usually lasts about four weeks. Trees vary in the amount of sap they will yield and in the percentage of sugar in the sap. The amount depends on the individual tree, and on the character of the soil and the situation. It fluctuates with night and day and with a change of temperature. Trees average about 12 gallons of sap a year or about 3 pounds of sugar. The amount may be much lower and individual trees have been known to yield 75 gallons of sap and to make 30¾ pounds of sugar in a single year. Sap usually contains about 3 per cent of sugar, but it may run as high as 10 per cent. Moderate tapping does not injure the trees or affect the growth or the quality of wood.

In tapping a tree the loose bark should be brushed away so that pieces will not fall into the pail. The hole is bored with a ⅜-inch or a ½-inch bit about an inch deep, directed slightly upward so as to insure drainage. Small holes are used so that they will close over in a year or two. The greater part of the sap comes from the outer 10 to 15 rings of the wood, so that deep boring is unnecessary. Wooden spouts are used, made from elder or sumach, the pith being forced out with a stick or burned out with a hot iron. The pails for collecting the sap should be covered to keep out bark, dirt, snow or rain. Cleanliness is necessary to prevent souring or spoiling of the sap.

The sap is usually boiled down in kettles in the woods or in sheds built for the purpose. As impurities rise they should be skimmed off. When the boiled down sap reaches a weight of 11 pounds to the gallon, good sirup will be produced. It should be dipped out and strained through flannel. It can be put up either hot or cold. It is believed by many better to can it cold and as soon as possible. For making sugar the sap should be boiled down until it begins to "sugar off." Old-fashioned methods to determine this point are by pouring the sirup on snow or by dipping in it a twig bent into a loop. If it becomes waxy on the snow or if it forms an elastic film on the loop, the sirup has boiled enough. It is then poured into moulds to harden. It should be stored in a cool, dry cellar or storeroom.

CHAPTER XII

PROTECTION OF THE WOODLOT

A WOODLOT must be protected from its enemies if it is to be successful. Trees are attacked by disease and insects the same as other forms of plant life, and they are subject to injury from fire, storm, animals, and the acts of man.

Tree Diseases.—There are many kinds of diseases to which trees are subject. The most common of them are caused by fungi. These are low forms of plant life that cannot manufacture food, being without chlorophyll, and so live on other plant life. They live on the wood and other parts of a tree, consuming the woody tissue and causing what is known as rot. The fungous disease is spread by spores that are very light and small and are carried through the air. Falling on the exposed surface of wood the spores will begin to grow, forcing their way into the tree and branching out in all directions through the woody tissue. After the fungous disease is well established in the tree it produces a fruiting body which is often in the shape of a shelf-like bracket or of a toadstool. Mushrooms and puffballs are the fruiting bodies of fungi. The fruiting bodies vary in the different kinds of fungi from a foot or more across to those barely visible. They vary in color also from colorless to black or red, but never green. From these fruiting bodies the spores are scattered by the wind.

As long as the bark on a tree is intact it will protect the tree from most kinds of fungi and decay. The spores may enter a tree through a broken limb and work down through the body of the tree, or they may enter through an injured

Fruiting bodies of rot producing fungi on the stump of an oak tree.

Photo by A. S. Rhodes.

Section across the butt of a chestnut tree, showing how decay has entered the tree through a fire scar.

root. Usually rot starts in a tree as the result of the breaking of the bark on the trunk by grazing animals, by accident or because the inner living bark has been killed by fire or other causes.

While most fungi grow on the stem and limbs of trees some attack the leaves, others the living bark and others the roots. Some fungi attack only the heartwood, in which case the trunk may become hollow. Since the heartwood has nothing to do with the life process of the tree, the wood being dead, heart rot does not often kill a tree. Trees are often seen growing vigorously with a hollow trunk. Heart rot destroys the value of the tree for lumber or other useful purpose and so weakens the tree that it is liable to be broken off by a heavy wind. Some fungi attack the sapwood and the cambium, the living layer of the tree. These fungi are most dangerous to the life of trees and soon bring about their death. The chestnut bark disease, which threatens the destruction of all chestnut trees, is of this character. It enters the bark through a wound or even through the lenticles of the bark. It spreads out and develops in the cambium layer, forming a spot that gradually grows larger and larger until the tree is girdled and killed. From the brown patches which it forms on the surface of the bark the spores are spread. The recently imported white pine blister rust is of much the same character.

Little can be done when trees are attacked by fungi, especially in a large woodlot. Because the disease is inside the bark, spraying will have no effect as it will not reach the disease. Injections of so-called remedies, or boring holes in a tree and filling with medicine or chemicals of different kinds will be of little or no value. The bark on trees in the woodlot should be kept intact. Wounds caused by pruning off branches or otherwise should be painted with a good paint or tar to prevent the spores or seeds of the fungi entering the wood. Fire that kills the inner bark at the base of trees should not be allowed to run through the woodlot and grazing animals should be excluded. Trees and limbs attacked by the chestnut bark disease or other diseases should be cut and the bark removed and burned. This will often prevent the spread of the disease to other portions of the tree or to other trees. Healthy, vigorously growing trees are less apt to be attacked by rot because new wood closes over wounds quickly. Anything that tends to induce health and vigor in the trees will help to prevent decay.

Insects.—Insects are another cause of damage and often of great loss to the woodlot. Every tree has its insect enemies that live and prey on it every year. Ordinarily they do but little damage. It is only when an insect becomes very numerous and it begins to do considerable damage that our attention is called to it. The destruction of birds is said to have been one of the chief causes of the increase in insect pests in recent years. So numerous have the insect pests of trees become in some regions that trees can be grown only by constantly fighting their insect enemies, by spraying the leaves with poison, destroying the eggs of insects, burning their nests, and in other ways.

Insects kill trees by defoliating them or eating the leaves, by sucking the juices through the leaves or bark, or by girdling the inner living bark. Some insects live only in the dead wood or heartwood of a tree or only on dead trees, while others attack only living trees. Some live only on one tree or a few trees, while others live on many kinds.

It is only by becoming acquainted with the life history of insects that they can be combated intelligently and successfully. Insects that spend the winter in webs in the trees like the brown tail moth or fall web worm can be destroyed by cutting off the webs and burning them or by burning the webs with a long-handled torch. Each web of the brown tail moth will contain from 500 to 900 caterpillars. The gipsy moth lays its eggs in yellowish clusters in crevices in the bark, along fences or in other places. These egg masses each containing 400 to 500 eggs can be destroyed by wetting them with creosote. Both of these insects have done great damage to the woodlots in New England and are likely to spread. The white pine weevil which attacks the tip shoots of the white pine trees, especially the young trees in plantations, can be killed by cutting off the infested leaders in the late summer and burning them. The weevil lays its eggs on the leader and the young develop in the growing twig. Leaders that are attacked begin to change color in the late summer. They should be cut at once and destroyed before the weevils escape. Caterpillars can be prevented from climbing trees by binding a band of sticky fly paper around the trunk of the tree. Insects that eat the leaves of trees can be destroyed by spraying the leaves with a poison solution, such as 10 pounds of arsenate of lead dissolved in 100 gallons of water. Insects that suck the juices from leaves and bark can be killed by a contact poison which will smother them, such as kerosene emulsion, soap and water, tobacco extract, or lime and sulphur wash. When trees become badly infested with insects that have bored beneath the bark, the trees should be cut and the bark burned. Boring insects can be cut out with a knife where an especially valuable tree is attacked, or carbon bisulphide can be injected into the burrow and the opening immediately plugged with putty or soap.

Dead trees and rotting branches and stumps are often the breeding places of insects that later may attack living trees. The woodlot should be kept free of such material by utilizing it for cordwood. The tops of trees cut should be utilized in this way or piled and burned. Many methods for combating insects can be carried out only on the small woodlot. They would not be practical with large forest areas. Where insects are destroying large portions of forests the only practical remedy is to cut all the trees in the infested area.

It is sometimes possible to combat an insect attack by

cutting from the woodlot the trees on which the particular insect feeds and encouraging the more resistant kinds. In starting a woodlot by planting only those trees should be used that are not subject to insect attacks. It is impossible to grow the black locust in many parts of its range because of the attacks of the locust borer that riddles the wood of the trees. It is said that this insect lives on the pollen of the goldenrod and that where this is destroyed locust trees can be grown. There is no practical method of combating many kinds of insects.

Fire.—Fire is one of the worst enemies of the woodlot. It not only injures large trees but it destroys the young trees and seedlings on which the future of the woodlot depends, and also destroys the favorable conditions of the forest floor and soil. It is often thought that a surface fire passing through a woodlot consuming the fallen leaves does scarcely any damage to the larger trees. This is because the crowns of the trees are elevated high above the fire and the bark on old trees is usually thick. A surface fire may be hot enough to scorch and kill the living cambium layer lying just beneath the bark without the tree showing scarcely any outward evidence of injury. The damage done to trees after a fire often does not become evident for many years. When the cambium layer is killed by being scorched, growth will cease over that portion of the trunk of the tree and decay of the wood will soon set in. The cause of so many large trees being rotten at the base is usually surface fires that have killed the cambium layer. Fungi gaining entrance through the dead bark spread out in the wood often rotting the heartwood for long distances. When such trees are cut they must be butted off many feet to get above the decay. On level land trees are injured most from a surface fire on the leeward side or the side away from the direction from which the fire comes. The direction in which a fire burned and usually the place where the fire started can often be determined by this means.

The effect of fire and wind on a stand of western white pine. The trees are all dead. The humus and the roots of the trees have been burned, causing the trees to be blown over by the wind.

Fire not only injures or destroys the trees; it burns off the humus that may have taken centuries to form.

The injury to the trees by surface fires can be shown immediately after the fire by removing the outer bark. If the inner bark looks brown or scorched the fire has killed the cambium or growing layer at that point. A fire in a woodlot always does great damage to the surface roots of trees.

It is sometimes held that a fire is an advantage to a young stand of growing timber, in that it will kill a few of the trees and the rest will grow all the faster because of the greater room in which to spread their crowns. A thinning is often needed in a thick stand of young trees, but fire does not exercise an intelligent choice as to the trees to be removed. The trees killed are those most susceptible to damage by fire, as those with thin bark or that contain a great amount of water. It may happen that these trees are of the better kinds and that those that will withstand fire the best are the poorest kinds of trees in the woodlot. Fire passing through the woodlot will often in this way bring about a change in the species and cause the woodlot to deteriorate in character.

It is in the destruction of the young seedlings and saplings and of the seeds lying in the humus ready to germinate that the greatest injury is done to the farm woodlot. The woodlot can get along without the older trees. It is natural for them finally through old age or disease to die and fall to the ground. But a woodlot cannot get along without the younger trees. As has been said, these are the children of the woodlot, and if they are continually killed off by fire or other causes there can be no future to the life of the woodlot.

Fire passing repeatedly over the ground destroys the humus and breaks down the crumbly character of the soil, making it hard, much like a pavement. This prevents the circulation of the air in the soil, which is necessary for root development. It tends to prevent the entrance of moisture, the water running over the surface instead of sinking into the ground. It prevents the new tender roots that start from germinating seeds from entering the soil, and so prevents the reproduction of the trees. The destruction of the humus or the decayed leaves and twigs lessens the moisture holding capacity of the soil. Such conditions are not favorable for the growth of trees, but encourage the growth of weeds.

Protecting the Woodlot from Fire.—Where woodlots are portions of larger forest areas some means must be taken to prevent fires spreading to them. Woodlots that are isolated, being surrounded by cultivated fields, are not usually subject to fires. If fire starts it is due to carelessness in most cases. The watchfulness of the owner is the surest safeguard against fires. If a fire can be discovered early before it has become of large size it usually can be extinguished easily. Where it is possible to do so there is no better means of protecting a woodlot than to run a few furrows around it with a plow. This will stop an ordinary, slowly moving surface fire. It also gives a base from which to back fire against a rapidly approaching fire. In back firing, a fire is lighted along a prepared line, path or road running at right angles across an approaching fire, and is allowed to burn back against the fire, consuming the inflammable material. This is a most effective means of combating a rapidly moving and hot fire. Fire lanes are sometimes made by turning furrows several yards or rods apart and burning off the inflammable material between. Such fire lanes are comparatively inexpensive and are very effective. They can be run along railroads and the material between the tracks and fire lane burned. They can be constructed in the same way along a highway where there is danger of fire being started accidentally. It is not always possible to plow fire lines owing to the nature of the soil. In this case lines can be raked or hoed clear of all inflammable material such as leaves, twigs, weeds and grass, and all branches and fallen trees removed. Regular roads and existing logging roads, trails and streams can also be utilized as fire lines.

Woodlots can be divided by fire lines into blocks. If fire starts in one block the lines will prevent it spreading to other portions. Farmers often delay too long making provision for preventing fire. A fire line ceases to be a fire line when it becomes covered with dry leaves and other material. Even in that condition they will serve, however, as trails along which men can move in fighting a fire. Experienced men often will not enter the woods to fight fire unless along a trail that will also lead them out of the woods in case of danger. Lines should be opened up through large woods for this purpose. The ground in a woodlot should be kept clear of all inflammable material. Dead tops and fallen limbs as well as the slash after logging should be worked up into cordwood and the rest piled and burned.

A surface fire passing through a woodlot often can be beaten out by the use of wet sacks or branches of trees. Where sand or dirt can be had it can be thrown from a shovel on a fire with good effect. Water where it can be obtained is probably the best means of fighting fires in the woodlot. A pail of water and an ordinary force pump will put out many feet of fire or lay it so low that it can easily be beaten out. In some regions where woodlots are of value fire extinguishers are used and work well. In thickly settled regions fire engines and trucks equipped with fire fighting tools are held in readiness to go to woodlot fires and extinguish them.

Where fire burns down below the surface and runs in dry humus where it has accumulated in thick layers, trenches must be dug down to the mineral soil around the fire to keep it from spreading. Such a fire is called a ground fire. It burns slowly destroying the humus and soil and causing great damage by burning the roots of trees causing them to fall over in the first wind.

The time to fight a large fire is at night when the dew has dampened the humus and foliage and the fire burns low and slowly. Under the drying and warming action of the sun the woods become dry and wind is apt to rise or change suddenly. A fire in the daytime usually burns too hot and too fast to be attacked by beating it out. A fire in a small woodlot can usually be stopped by raking the leaves and other dead material from in front of the fire or by beating out the flames.

Grazing.—In a mature, even aged woodlot where the trees are large and the canopy is far removed from the heads of cattle and where reproduction of the trees is not looked for or desired, the grazing of a few head of cattle can do but little immediate harm to the standing trees. It is overgrazing that does harm. But in a recently established forest planting, or in a young growing woodlot, or in the ordinary farm woodlot composed of trees of all sizes and ages, or where reproduction of the trees in the woodlot is desired, grazing must be excluded.

Animals do damage in several ways when grazing in a woodlot. The result of their action is much similar to that produced by fire. The young seedlings and saplings are eaten and destroyed, the ground is packed hard by the constant trampling of the cattle, making it impervious to water; the surface roots of trees are exposed and injured, and the bark on larger trees is rubbed off or gnawed exposing the wood to decay. Different kinds of animals affect the woodlot differently. Sheep cut into the soil with their sharp hoofs, pulverizing it and injuring the surface roots of the trees and seedlings. Horses and cattle destroy large numbers of young trees by trampling. Damage is done to the trees by browsing on the young growth, by biting off buds, leaves and shoots, and by gnawing and rubbing off the bark. Horses reach higher and are fonder of leaves than cattle, and because of their iron shoes they do greater damage. Sheep have their heads low, but they eat close and are exceedingly harmful to the woodlot. Goats prefer the leaves and bark of trees to other forage. They should never be allowed in the woodlot. Goats and sheep are often used to clean up brush land because of the damage they do to young trees and sprouts. From 3 to 5 goats to the acre will usually clear out all the tree growth in a single season. Hogs root up the soil and eat the young succulent roots of trees. They are sometimes driven into the woodlot to expose the mineral soil so that the seed from the trees will germinate better.

There is much truth in the old saying that it is impossible to grow trees and animals on the same ground. Where it is necessary to run cattle in the woodlot for protection from the hot sun in summer and from the cold winds of winter, the woodlot should be divided, a portion being fenced off and devoted to this purpose and the rest of the woodlot used for growing trees.

Storms.—Considerable damage is done to trees in the woodlot by storms causing windfall and breakage. Trees in the woodlot protect each other from the wind. Standing closely together, the crowns prevent the wind blowing the trees over so far that they will be uprooted or broken off. Great care must be used in thinning a woodlot not to suddenly remove so many trees in one spot that the tree crowns will be separated from one another so far that windfall or breakage will result. Windfall is especially likely to occur in moist situations, where the trees naturally grow shallow root systems. It is well to maintain a mantle or protecting hedge to the woodlot to keep the wind from getting in under the trees. Trees on the outer edges of the woodlot should be allowed to grow branches low down on the trunks, and shrubs and smaller trees should be encouraged to grow. It is a mistake to open up the edge of a woodlot by pruning up the trees and by removing the young growth and shrubs found. Such a protective mantle prevents the cold winds of winter from injuring the trees by freezing and cracking open the trunks. It prevents the hot winds of summer from drying up the soil under the trees and also helps to prevent windfall and breakage.

The Owner.—Probably the worst enemy of the woodlot

is the owner himself. By his lack of judgment as to the trees to cut for fuel, always selecting the most vigorously growing and best formed trees and leaving the poor ones, the woodlot becomes filled with old, crooked and misshapen trees. By giving no thought to the trees that start to grow in the space left by the removal of a tree, allowing nature to fill in the opening with any tree that happens to come up, the woodlot often becomes filled with weed trees that have little use on the farm. Only by the removal of the weed trees and by favoring and planting trees of good species can an owner secure trees in the woodlot that will be of use. Many woodlots after years of cutting are in a culled condition, all the valuable species having been removed. The kinds of trees that could not be sold were left to sow the seed for the future woodlot. Many woodlots formerly filled with useful and valuable trees now contain only inferior kinds of trees. The surest way to secure valuable trees in the woodlot is to plant them whenever a tree is removed.

By not growing his trees closely enough together the trees become limby and grow short trunks. By not thinning out the trees as they gradually grow larger he permits the trees to expend most of their energy in fighting neighboring trees rather than in rapid height and diameter growth. By giving to each tree the amount of light and space it needs by frequent and light thinnings, removing the weaker and inferior species the owner can greatly increase the annual yield of the woodlot. Too many farmers leave the growing of their trees to nature. Nature's methods are usually not profitable and in the woodlot she will often grow but one-fourth or less of the amount of wood product that can be grown by proper care and attention. By not making provision for trees to grow up to take the place of those removed, breaks occur in the canopy causing a loss in yield. By not encouraging the fastest growing species that will serve his purpose, he produces annually but a small part of the wood product he might

A protective mantle to the woodlot protects the trees from cold and dry wind.

Grazing is injurious to the trees in the woodlot. Dead limbs in the tops of the trees show lack of vigor.

obtain. By not protecting the woodlot from fire and grazing he is destroying the future of the woodlot and injuring the present stand of trees. Insects and fungi, because of his neglect, use up a large percentage of the product.

As has been said, a woodlot will respond to good treatment and the owner can by giving a little attention to the growing trees, and protecting them from their enemies, make a woodlot a source of income as well as have it furnish the farm with all the wood desired.

CHAPTER XIII

WOODLOT MANAGEMENT

Conception of the Woodlot as Forest Capital.—The trees in the woodlot can be considered as forest capital from which an income is to be derived or the same as money in the bank at interest. The amount of wood that is laid on the trees in a year, the annual growth, is the interest on the forest capital. If the amount of wood produced by each tree every year is large, owing to rapid growth, a high rate of interest will be earned by the woodlot.

Unless the trees in the woodlot are fast growing species of trees and unless they are given the proper amount of space for their best and most rapid development by means of thinnings, a low rate of interest must be expected. Like any other business, the returns from the woodlot will depend on the kind and amount of capital in the business, and also on how that capital is handled.

The Amount of Timber to Remove Annually from the Woodlot.—It is important to know the amount of timber that can be removed annually from the woodlot without detriment to the forest capital. It is a common saying that a farmer cannot use his woodlot and have it too. This is because as soon as he begins cutting in the woodlot it is not many years before the woodlot is in a wornout condition, the trees standing far apart and grass and weeds covering the forest floor. This results from not knowing the amount that can safely be removed each year. In this case instead of cutting the interest only, the owner has been cutting into the forest capital. The amount that can safely be cut from the woodlot each year

This kind of woodlot can yield but a small annual income.

is the amount that grows each year or the interest earned by the forest capital. If more is cut the woodlot will soon run out.

A cord per acre per annum is an old measure of the growth of the ordinary woodlot. There are, however, few woodlots that grow at that rate. Most of them produce but half a cord or less per acre every year. In cutting it will not be safe to figure on a high yield. Raising the yield is the problem of the farm woodlot. With a well stocked stand of fast growing trees that are given the proper room for their best growth by means of thinnings, a cord per annum and more should be produced. The yield should be placed at a certain amount and if the owner finds that after cutting a few years the woodlot is becoming thin, a less amount should be removed annually. If, on the other hand, the woodlot is becoming overstocked with large trees, a larger amount can safely be removed.

Size of the Woodlot.—Whether a woodlot will supply a farm with all the wood products needed will depend primarily on the size of the woodlot. It is often said that one-eighth of the farm should be in a woodlot and that area should supply the farm for all time with all wood products needed, and also furnish some material for sale. The area that should be in farm woodlot, however, is not definite. The annual income from an acre of wooded land will depend on the species, whether it has slow or fast growth; on the soil, whether it is suited to the tree grown; on the moisture content of the soil, whether it is wet or dry; on the aspect or exposure to winds or storms, and on the topography, whether the ground is level or a steep slope, and on which way it faces. All these factors and others will affect the growth of trees. Even the same tree will not grow at the same rate in the same region if the character of the soil varies or the moisture in the soil differs in amount. The annual growth must be determined for each particular woodlot and the size figured accordingly.

If the woodlot is found to be too small to supply the farm with all wood products needed its boundaries should be extended by planting trees, or wood products should be purchased.

Regulating the Cuttings in the Woodlot so as to Secure an Equal Annual Yield.—One of the objects of woodlot management is to so arrange or regulate the cuttings that an equal amount of wood can be removed from the woodlot year after year or periodically without lessening the forest capital. In other words, to put the woodlot on a business basis with a sustained annual yield.

One of the means of accomplishing this object in a woodlot where the stand of timber is fairly uniform throughout is by the area method. This consists in dividing the woodlot into as many equal areas as the trees are years old when mature and cutting one part each year. The age at which the trees are cut is known as the rotation. The rotation will vary with the product desired, whether it be sawlogs, poles, ties, posts or cordwood. Even with the same product the age at which the trees will be large enough to cut for a certain product will vary with the rate of growth of the trees. The faster the trees grow the lower the rotation can be made.

The age at which the trees are to be cut having been established, the woodlot is divided into that number of equal parts and the timber on one of them is cut each year. As soon as the trees have been cut reproduction is established by seedlings or from sprouts and a new stand of timber started. In this way by cutting one area a year the entire woodlot is finally cut over and the cut-over areas reproduced. At the end of the rotation or the time it takes to cut through the woodlot, the trees on the first area cut will be old enough and large enough to harvest. The owner starts and cuts through the woodlot again. Each year the timber on one area becomes mature and is cut and reproduction is established.*

* The rotation can also be divided into periods of 5 or 10 years

In this way the woodlot is placed on a financial basis the same as any other business. It produces a known income every year. The value of the woodlot can be figured on a basis of the income.

This represents the method of management of forests in countries where forestry has been practiced many years. It gives an idea of what forestry really is, a business of growing crops of trees. There are many other methods of regulating the annual cut in woodlots or forests to secure a sustained yield. These can be found described in books relating to general forestry.

TAXATION OF WOODLOTS

Woodlots in most states are subject to the general property tax. Each year the assessors appraise the value of the trees together with the land, and the owner pays the tax annually on this valuation. No one objects to paying a just tax on property, but a tax out of proportion to the value of the property is unjust. A mature stand of timber that is being held for a rise in the value of stumpage is probably justly taxed by this system. But often the tax on such timber has been made a burden. Overtaxation on mature timber has been common in many regions, the assessors figuring to get all out of the timber possible before it is cut. This has been a chief cause of the slashing of many of our forests.

But a growing stand of timber, on the other hand, is in the nature of a crop and should pay a tax only when the crop is harvested, and not year after year while the crop is growing and producing no revenue. When the crop brings in an income then the tax should be collected. The land should pay

each and the woodlot divided into as many parts as there are periods in the rotation. Then the timber on one part would be cut during each period and reproduction established. When the woodlot is cut through the timber on the first part cut would be large enough to harvest.

its just tax each year on the basis of its productive capacity the same as plowland, and the timber should pay its tax when cut. The paying of a tax each year during the life of a growing crop of trees makes the total tax paid out at the time the timber is cut, out of all proportion to the actual value of the timber, especially since the money paid out each year must run at compound interest up to the time the crop is harvested.

Many of the states have given some attention to the taxation of growing timber. In some of them legislation has been passed by which growing timber is placed in a separate classification for purposes of taxation. The plans differ in many respects, but all advocate that the land should be assessed on a basis of its earning capacity and that the timber should pay a yield tax when harvested. In New York State the timber when harvested pays a yield tax of 5 per cent of the value of the crop. Michigan lays a tax of 5 per cent on forest plantations when mature, Massachusetts 6 per cent, and Vermont, Connecticut and Pennsylvania 10 per cent, with different provisions for forests already established. In Pennsylvania the land must be classified by the State Forestry Department as auxiliary forest reserves, and thereafter the land cannot be assessed for purposes of taxation at a higher valuation than $1 per acre. The state compensates school and road districts for the loss of revenue by paying into the local funds for each of these purposes state moneys at the rate of 2 cents per acre on land so classified. Vermont, Massachusetts and Connecticut provide for a valuation of the land separately from the timber and when this valuation has been made it must stand for a long term of years. There are many variations in the details of the laws, but they all agree in embodying the principle of a tax on the yield separate from a tax on the land. In some states reforested lands are exempted from the payment of taxes for a certain number of years, while in others bounties and rebates are given.

Damages.—When timber in a woodlot is injured by fire

or other causes damages can be collected. The usual measure of damages in such cases is the difference between the value of the timber before the damage and the value after the damage. The valuation of trees in woodlots for the purpose of determining damages is usually based on the amount of cordwood it will cut. Before the courts generally if the trees are not large enough to cut for cordwood they have no value. A future value of young forest trees is not recognized.

If a young apple orchard is burned soon after it is set out and before it has begun to bear fruit, damages can be collected far in advance of the cost of the establishment of the orchard plus the interest and taxes and other expenses to date. The courts recognize a value in a young orchard that is based on the expectation of the owner of harvesting crops of apples in the future. This value is sometimes called the expectation value. A young colt has but little value in itself. It does no work and is a constant expense. If killed the owner can, however, collect damages based on the future value and usefulness of the animal. The same reasoning does not yet apply to young stands of timber. A value based on the future returns or the expectation value is not recognized. The expectation value would be the future sale value discounted back to the time of the fire minus any future expenses. There is no reason why such a value should not be recognized and allowed. That it is not is due to the fact that the principles involved are not thoroughly understood.

In addition to the damage to the present stand of trees and to the future growth, there is also a damage to the soil resulting from the burning of the humus that has been many years accumulating and which acts like a fertilizer to the trees. Future crops will grow much slower because of the loss of humus in the soil.

Often damages to trees do not become evident at once. After a winter or spring fire in which trees have been injured badly so that they have been completely girdled, they will often leaf out in the spring and grow vigorously and there will be no evidence of serious damage having been done. The same trees will die in midsummer. They leaf out in the spring owing to the large amount of food material and moisture stored in the trunks of the trees. As soon as the stored food has been used up and the moisture withdrawn the trees will die. If the bark is peeled away from the lower part of a tree after a fire the damage, if any, can be shown by the scorched color of the cambium layer or inner bark. Trees only partly girdled are badly damaged because of the death of the cambium layer, which subjects the wood to attacks by fungi and insects that may render a large portion of the tree unfit for use.

CHAPTER XIV

THE PRODUCTS OF THE WOODLOT

THE chief products of the woodlot, besides material used on the farm, are sawlogs, lumber, poles, ties, posts, mine timbers, cordwood, etc.

Sawlogs.—There is usually a good demand for sawlogs by local sawmills. The woodlot owner should know what kinds of logs and what lengths can be marketed and the value of such logs. Often the butt log, as with hickory, is worth more than the upper logs. A special market should be sought wherever possible. Perfectly sound logs are in demand for many purposes. Logs of black walnut are in demand for gun stocks, black cherry for furniture, ash for tool handles, willow for gunpowder, white oak for quarter-sawed stock, tulip poplar and others for veneers. Nearly every kind of wood has a use for which it is adapted, and for which a high price is paid for good stock.

The standard lengths of logs are 12, 14 and 16 feet. Short logs are often cut 8 and 10 feet long, and logs longer than 16 feet for special purposes. It is the rule to cut logs a few inches longer than the length of material to be manufactured, to allow for bruising the ends in skidding the logs out of the woods and also for squaring the ends of lumber. Logs should not be left long in the woods after being cut. If it is not possible to saw them into lumber at once or deliver them to the mill they should be skidded out to an open area and piled on skidways. Painting the ends of logs or coating with tar will prevent checking or splitting of the ends.

Lumber.—Lumber is cut from nearly all species of trees

and has a value according to the species and quality. A good local market is often found among neighboring farms and small towns. Special orders for local house or barn construction can be filled. The product of local sawmills is not in demand in general competition with the product from large sawmills, because most of the product manufactured is of poor quality. Because of the light weight of the sawmill machinery and lack of skill on the part of the sawyer in keeping his mill in condition and in the sawing of the lumber, the material is not uniform. One board or stick may be a little wider or thinner than another. Boards are often sawed thicker on one end than on the other. Carpenters like to use a product that is uniform throughout and that does not vary. The product of large mills is in demand for this reason. Yet many portable sawmills turn out an almost perfect product.

Boards sawed from the same log do not all have the same value. Those sawed from the outer portions of the log will usually be clear, without defects or knots. The farther into the center of a log a board is cut the more knots and imperfections it may have. Boards and other lumber is graded according to the number, size and location of the defects contained. The more perfect the lumber and the freer from defects the higher the grade and the higher the price the lumber will bring. Lumber is sometimes sold as it comes from the mill, all grades mixed in together, called the mill run. Where lumber is to be sold by grades a knowledge of grades is necessary. Often an experienced grader can be employed, if not the owner should make a study of the chief grades of different kinds of lumber so as to be able to sort the boards. Standard grades for lumber have been established and are in use and can be secured through the State Forester. The lower grades have but little or no market value. They are called culls. These can be used about the farm as needed or cut up for firewood and kindling. More can usually be made by sorting the boards according to the defects and kinds and selling by grades than by selling the mill run.

The value of lumber will also vary according to the way it is sawed from the log. Boards cut from the side of a log have what is called flat grain. The annual rings of wood will pass through the board at a greater angle than 45 degrees. Boards cut running from the edge toward the center of the log will have the annual rings more or less vertical to the surface of the boards. Such logs are called rift-sawed or quarter-sawed boards. In quarter-sawed boards the medullary rays will pass through the board from edge to edge. With woods that have large medullary rays, as white and red oak, cherry and sycamore, the medullary rays where they come to the surface of the board and are cut across by the saw, will give the figured grain which is in so great demand for furniture and decorative purposes. The name quarter-sawed boards comes from the early practice of first cutting the log into quarters before the quarter-sawed stock was cut. Quarter-sawed boards shrink less and wear better than flat grained boards, and so are of especial value for flooring.

It is often necessary to keep lumber several weeks or months after it has been sawed before a sale is made or for the purpose of seasoning the wood. The lumber should be sorted as nearly as possible by grades, and the different kinds and sizes piled by themselves neatly in the lumber yard. This should be located in an open area where there is a good circulation of air and good drainage. The lumber should not be piled on the ground, but should be raised a foot or more on crosspieces. The edges of the boards should not touch each other and between every two tiers of boards well-seasoned slats should be placed every few feet to permit the air to circulate freely. The front and rear cleats should be laid flush with the ends of the boards to prevent checking, and cleats should be laid one above another to keep the boards from bending. With woods that have a tendency to warp the cleats should be placed every 2 or 3 feet, and with woods like pine and spruce they can be placed much wider apart.

Poles.—There is a demand for poles in nearly all regions for telephone and telegraph lines and by street railroads and electric power lines. The kinds of wood used for poles will vary in different regions. Usually a strong, light and soft wood is demanded, but use is made of the species that grow in any region, for it is not practical to ship poles any great distance because of their weight.

The price paid for poles will vary with the species. The value of poles of the same kind of wood will depend largely on the length and condition. Often it will be found possible to secure a greater profit from poles than from sawlogs or ties. A farmer should be able to figure such problems. Poles should be cut according to specifications. Poles are usually classified by 5 foot lengths beginning at 20 feet and running up to 60 or more feet, with a top diameter of 7 inches for the smallest poles. There exists in many regions a good market also for piles. They are used for foundations for wharves, quays, bridges, retaining walls, etc. Converter poles find a ready market near smelter plants.

Poles and piles should be cut only when there is a market at hand, for they deteriorate rapidly. If poles cannot be marketed they should be peeled and skidded out of the woods to an open spot where the air circulates freely and rolled on skids so that they will be well off the ground. Poles should not be piled one on top of the other, but should be one layer deep, and they should be laid several inches apart to insure good seasoning.

Cross-Ties.—Where there is a demand for ties they are a valuable product of the woodlot. They are either hewn out in the woods or the logs are hauled to the sawmill and sawed to proper form and size. There are several grades and sizes of ties, from the highest grades used in the main lines of railroads to the small and short ties used in mines. A woodlot owner should have in mind the grades and sizes of ties for which there is a market in his region and manufacture them accordingly. Formerly only hard, durable woods were used for ties, such as white oak, locust, etc. Owing to the growing scarcity of the more durable woods many other kinds of woods are in demand. The process of treating wood with creosote to prolong its life in contact with the ground has made possible the use of many perishable woods for ties, such as hard maple, beech and birch.

It may be possible, owing to the location of the woodlot or the market, that logs made into ties will yield far more than when sawed into lumber. Such facts can be determined only by careful investigation of the conditions. The aim should be in cutting the trees in the woodlot for sale to manufacture the products that will yield the greatest profit.

Fence Posts.—Fence posts are cut in different lengths according to the demands of the user. The standard post is 7 feet long. The size around will vary often with the use. Line posts, for example, need not be as strong as corner posts. Usually larger posts are in demand for it is believed that they will last longer in the ground before rotting off. Posts should be large enough so that they will have sufficient heartwood to furnish the strength desired when the sapwood decays, for the sapwood of all kinds of posts will decay quickly.

Many kinds of woods are used for fence posts. The length of life of posts will depend primarily on the durability of the wood. Some kinds of posts will last but two or three years in the ground before rotting off, such as hickory, soft maple and birch, while others will last many times as long, as locust, cedar and chestnut. The more durable the wood the greater the value of the posts. By giving fence posts a preservative treatment with creosote the most perishable woods can be made to last many times as long as they ordinarily would. This process is described in a subsequent chapter. Posts should not be left lying on the ground after being cut, but should be set on end or piled in open piles so that they will season well.

Mine Timbers.—In the neighborhood of mines there is usually a good demand for mine timbers. These vary from heavy props and supporting timbers to small sticks but 3 or 4 inches in diameter called lagging, used to keep earth and rock from falling into the mine between the props. Many mine ties are also used. These are much smaller in size than the ordinary ties used in railroads. Wood is used in mines in preference to other material, because it will give way slowly when subject to great weight, gradually splintering and cracking, giving the miners warning of the moving rock and a chance to get away. Wood decays very rapidly in mines owing to the damp conditions so it must be replaced often. Durable woods are in greatest demand but where these are scarce almost any strong wood is accepted.

Cordwood.—Fuel wood is usually cut into 4-foot lengths and piled in stacks 4 feet high and 8 feet long. Such a stack occupies 128 cubic feet. A cord foot is one-eighth of a cord or a pile 4 feet high, 4 feet wide and 1 foot long. Surface feet as applied to cordwood is the number of square feet measured on the side of a stack. Pulpwood is sometimes cut in 5-foot lengths, in which case a stack 4 feet high and 8 feet long will contain 160 cubic feet. Wood is also cut in shorter length, as 15 or 18 inches for stove wood, or 2 to 3 feet long for other purposes. A cord of such wood is a stack 4 feet high and 8 feet long. It is often called a short cord. In selling a short cord the price is made proportional to the length of the pieces.

The amount of solid wood in a cord will vary with many factors. The more important ones are as follows:

The Form of the Sticks.—If the wood is straight and smooth, such as body wood or trunk wood, the pieces will lie closer and a cord will contain more wood than if the pieces are crooked, branched or forked, such as limb wood.

Cordwood is one of the chief products of the woodlot.

The Method of Stacking.—Where the wood is thrown on a pile without much order, some pieces being crossed, less wood will be contained in a pile than if the pieces are regularly stacked. In measuring stacked cordwood it is customary to deduct a certain amount from the measurement if the wood is not carefully piled.

Whether the Wood is Large or Small.—The smaller the pieces of wood the more space they will occupy. A cord of wood swells by splitting the pieces. Large wood stacks closer than small wood. It is customary to split large pieces into smaller ones for fuel wood. This permits of better drying of the wood.

In general a standard cord will contain about 90 cubic feet of solid wood. Large, smooth, straight wood, such as pulpwood, will contain often as high as 100 cubic feet per cord, while a pile of small crooked wood may contain but 30 to 40 cubic feet or less.

Cordwood for sale should be piled neatly and regularly. It should not be stacked in the woods but should be hauled out to an open place and stacked where the sun and wind will quickly season the wood. It should be stacked on sticks run lengthwise of the pile and not on the ground. Rank vegetation should not be permitted to grow up around it, for it will keep the lower part of the stack moist and subject to decay. Cordwood is supposed to be made of green wood, cut, piled and seasoned. If too much dead wood is mixed in with the green wood the cordwood will have a dull appearance and will bring a lower price.

Very often special markets can be found for cordwood, especially near large towns. Selected pieces of hickory or oak cut to proper length for burning in fireplaces and well seasoned will bring more than the same wood cut and stacked in the usual way. Many wood-using industries purchase wood of certain sizes cut and piled in cords. Woods vary in the amount of heat given off in burning. In general the heavier the wood the more heat it will give out.

Other Products.—There are many other products furnished by farm woodlots most of them however, supplying only a local demand. More can often be made by supplying these special markets than from the ordinary products. In the vicinity of pulp mills soft wood like spruce, poplar, aspen, hemlock and pine are usually in demand. Acid factories and distillation plants require large quantities of wood like beech, birch and maple. Wood is in demand in many regions for wagon stock, tight and slack cooperage, and for making furniture and chairs. The bark of trees especially of hemlock and chestnut oak is in demand for tanning leather. Sometimes more can be made out of the bark of a tree than out of the wood.

The results of destructive lumbering. The future of the forest is destroyed.

The results of conservative lumbering. Many older trees left to sow the seed for a second crop. Young trees and seedlings protected. The future of the forest is assured.

CHAPTER XV

HARVESTING THE PRODUCTS OF THE WOODLOT

The object of maintaining a woodlot is the utilization of the wood products. It is seldom that a woodlot is grown for sentimental or esthetic purposes alone. There is no reason, however, why if properly cared for a woodlot cannot be ornamental as well as serve a useful purpose. A woodlot is for use primarily. When a tree is ready for the ax it should be cut. Forestry does not mean preserving trees, it means preserving the life of the forest growth. As we have already learned, forestry is the growing of successive crops of trees. This implies the harvesting of the crop when it is ripe. In forestry the trees would be cut when ripe the same as in lumbering; but forestry differs from destructive lumbering in that in cutting the timber the future life of the forest is provided for. In destructive lumbering the trees would be cut without any thought for the future, the object being to convert the standing trees into lumber and money in the quickest time and cheapest way. The result is generally a barren waste, due to fire and the destruction of young trees and seedlings. According to forestry methods, on the other hand, the trees would be cut not only to utilize them to the best advantage, but also with the future of the forest in view. The trees would be cut so that a new crop would spring up after the old trees are removed.

THE AGE AND SIZE AT WHICH TREES SHOULD BE CUT

The age and size at which a tree becomes ripe for the ax will depend on the purpose for which the tree was grown or the purpose for which it can best be utilized. If trees are grown for fence posts they are ripe and should be cut when they reach fence post size. If a fence post is wanted from a woodlot a tree suitable for the purpose should be selected and cut, provided it is not needed by the woodlot for protection to the soil or for seeding up the ground with seedlings, and provided it will not serve a better purpose later if allowed to grow larger, and also provided there are no poorer trees that will serve the purpose just as well. The aim should be to utilize each tree for that purpose for which it is best suited. Trees often are cut from the woodlot without any thought as to their fitness for any definite use. If a tree will make a good sawlog it should be kept until lumber is needed, and not cut for cordwood because the wood is straight grained and easy to split. If a tree has a wood that is suited for a particular purpose and for which there is a high demand it should not be cut for another purpose.

When trees are cut from the woodlot for sale they should be utilized when they will yield the highest returns in money. The size and age will vary in different regions with the conditions. Often trees cut when small and grown several short rotations by coppice will bring in a larger revenue than if held before cutting an equal time until they become large in size. Small second growth hickory, for example, is in demand for spokes and ax handles. It is often grown on short rotations for this purpose. As soon as the sprouts become large enough they are cut and sold. Chestnut and red oak coppice are grown on short rotations often of eight to twelve years to produce small sticks or lagging for use in mines. More can often be obtained in this way where there is a good market than if the trees were allowed to grow to large size before cutting.

As we have seen trees grow slowly at first, but soon begin a rapid height and diameter growth that continues to middle life or through the pole stage in some species, after that the rate of growth decreases until in an old tree the annual rings

Felling a tree with the crosscut saw.

Skidding logs out of the woods by horses.

are very narrow. In general trees should be harvested when they are financially ripe, or before the slower growth of old age begins. When the increased value of a larger product or its usefulness will make up for the loss due to the slower growth, trees should be allowed to grow to large size. When a tree becomes over mature in the woodlot because of the very slow growth wounds are not covered over quickly and the exposed wood is soon attacked by fungi and begins to decay. Trees in the woodlot should be harvested while the wood is sound.

Cutting Trees from the Woodlot.—The products from the small woodlot for farm use and for sale are usually harvested by the owner himself. He goes into the woods each winter with his help and cuts out such material as is needed. In cutting a tree an undercut is made with the ax on the side of the trunk, to determine the direction the tree is to fall and to keep the tree from splitting and kicking back. This should be cut about 2 inches deep in coniferous trees, and in hardwoods it should extend to the center of the tree. It should be made as low down on the trunk as possible. It is more difficult to saw a low stump than a high one, but much good wood is thrown away in cutting high stumps. The added material will more than repay any extra exertion necessary. Low stumps sprout better than high ones, and in growing trees from sprouts this is an important consideration.

The cross-cut saw is then used to saw the tree through from the opposite side. The bark should be chipped off round the tree where the saw will cut so that it will not impede the motion of the saw. The saw should be started about 2 inches above the notch and drawn back and forth with a regular, smooth swing. It should not be forced hard against the wood, but allowed to run freely in contact with it. If the kerf is sawed parallel to the notch the tree will fall in that direction. If one corner of the notch is sawed across before the other the tree will have a tendency to swing away from the severed corner. If the tree settles back and pinches the saw, wedges made from hardwood or iron are driven into the kerf back of the saw. This lifts the tree off the saw and helps to throw the tree in the desired direction. It is dangerous to attempt to throw a tree against a strong wind. As soon as the tree starts to move the saw should be drawn out of the kerf. There is danger in standing directly behind a tree that is falling, because it is apt to jump or kick back many feet. Standing near the stump one can readily judge the direction the tree is falling. It is safer to stand near the stump and to move back a few feet to avoid the jumping of the trunk than to run without looking. Often limbs broken off the tree or from neighboring trees will fall both before and after the tree is down, and should be looked out for.

After the tree is down the limbs are chopped off with the ax smooth with the trunk and the lengths of the products desired are marked off. This should be done with care so that the most possible can be made out of the tree. Usually the butt log in a sound tree contains the best material. If the butt is found to be decayed it must be cut or butted off. It is best to measure the entire length of the merchantable part of the trunk and then decide how it can be worked up to the best advantage. Defects can be sawed across so that they will come at the ends of logs. If the trunk has a crook it is possible to reduce its effect by cutting at the middle of the bend. A long log will usually bring a better price than two short logs of equal length. The lengths of the logs are determined usually by the market or the product desired. The whole trunk of the tree should be utilized and not part left in the woods. The upper log though full of knots can often be made into a tie or fence post. The products having been marked off they are cut by means of the cross-cut saw.

The logs must then be dragged or skidded out to the roads. This is done by horses, a chain being looped about the large end of the log or fastened to tongs. Often the edges of the log are beveled off or "nosed" to prevent them catching on stones or roots. Sometimes when there is snow the end of the log or of several logs are raised on a single sled and chained fast, the other ends dragging on the ground. The logs are skidded out to the road and either piled on skidways, from which they are rolled on to the wagons or sleds, or they are loaded from the ground. In this case two stout poles or skids are cut from the woods and laid against the side of the wagon or sled, and the logs rolled up them by means of a chain fastened to the wagon, passed under and over the log and back over the wagon to the horses. When the sled or wagon is loaded the logs are held in place by a chain and hauled out to the mill.

After the logs have been removed the rest of the tree should be manufactured into cordwood, fence posts or ties. All the tree of any value for sale or for use on the farm should be utilized in some way. Where there is danger from fire the branches should be piled and later on burned before the dry season begins.

CARE NEEDED IN CUTTING TREES FROM THE WOODLOT

The removal of a tree from the woodlot naturally causes considerable damage to surrounding trees and especially to the young trees and seedlings. Some damage cannot be helped, but it can be reduced to a minimum by proper care as to the direction the tree is to fall. With the use of a wedge a tree can be caused to fall in any desired direction, unless it is a badly leaning tree. It should be made to fall so that the crown will slip between the crowns of other trees so as not to break them. It should fall away from valuable young growth and if possible toward an open space. When a tree is to be felled it is usual to cut away all the young growth, seedlings and saplings, about the tree, so they will not interfere with the free swing of the ax. The ax is a dangerous instrument even to an experienced chopper, especially if it catches on a branch of a young tree and springs back. All precautions should be taken, but as little cutting should be done as possible. It is on these young trees and seedlings that start under the shade of the larger trees that the owner must rely to fill in the vacant spaces caused by the trees removed. Often a fine young sapling can be bent down and held in place by a forked stick driven into the ground until a tree has been cut. The sapling can be straightened up later without injury. The growth of many years can be destroyed thoughtlessly by a single slash with the ax.

In logging or skidding logs from the woodlot a little care will prevent useless breakage, bending and trampling of young growth. All the logs from a certain part of the woodlot can be dragged out over the same path often with much less effort and with less damage to the woodlot than where each log is skidded out over a separate path. When the trees in the woodlot are sold to lumbermen they should be made to protect young growth as much as possible. The owner should formulate and have carried out by written contract such rules in regard to the trees to cut and the methods of logging used as will tend to protect young trees and seedlings and benefit the future stand of timber. By this method it is possible to sell only certain material from the woodlot and to avoid the skinning of the land. The owner can have the defective and diseased trees and the trees of inferior species removed as well as those of high quality. If the inferior trees are left to grow and scatter seed and the better kinds cut, the future stand will be of less value than the present timber. The owner can stipulate that with the better kinds of trees those above a certain size only shall be cut, and that inferior trees shall be removed down to a low diameter limit. It is rarely that trees of less than 10 inches in diameter at breast height have any special value. They usually cut only a small amount of low grade lumber. The lumberman does not figure on paying

The portable sawmill outfit in the woodlot.

Hewing out ties in the woodlot.

much for them. He gets out of them whatever he can. Trees of that size, however, are growing rapidly, and will grow more rapidly after the trees around them have been removed opening up the crowns to the light. A lumber operation in a woodlot should be made to result ultimately in the establishment of a better stand of trees than the old timber.

Harvesting the Crop from a Large Woodlot.—When a woodlot is to be cut clean or when many trees are to be cut from a large woodlot there are two general methods that can be followed by the owner. The operation can be carried on by the owner himself or the timber can be sold to a lumberman and removed by him. Where the owner carries on the operation he can haul the product to a local sawmill and sell it or have it sawed into lumber; he can purchase a portable sawmill and do his own sawing, or he can have a portable mill brought on his tract to saw the timber. Which one of these several methods would be best to use could be determined only by a careful study of the conditions on each woodlot and of the market. If there is a good and ready market for the sale of all products from the woodlot, the owner can usually make more profit if he is a skillful operator by carrying on the operation himself and selling the products. If the market is dull and lumber and other products are hard to dispose of, it may be better to sell the stumpage outright to a lumberman who has made a study of markets and can usually dispose of the products at a profit. The advantage to the owner of carrying on the operation himself is that he can employ his help profitably throughout the winter, he can exercise care not to injure young growth unnecessarily or to break young trees or strip them of branches in felling other trees, he can leave the younger trees to grow to larger size, or he can leave seed trees to seed up vacant areas. It is not necessary for him to skin the land. He can cut and sell what he wishes and leave the rest. He can practice forestry in the cutting of the timber.

If a lumberman purchases the timber and carries on the logging operation he will probably give little thought to the future of the woodlot. His efforts will be directed toward removing the trees and young timber at the least possible expense. Having purchased the entire tract he will cut out everything that will yield him a profit. Many woodlots never recover from the destructive methods often practiced by lumber operators.

Lumbering, however, is a business in itself and requires expensive machinery, skill and efficiency in operating, and ability to dispose of the products. With most woodlot owners it will usually be best to dispose of the timber on the stump to a practical lumberman, making such stipulations in the contract as are necessary to protect the future of the woodlot.

COST OF HARVESTING PRODUCTS FROM THE WOODLOT

No definite figures can be given as to the cost of harvesting the products from the woodlot. There are many factors that will affect the cost of the different parts of an operation. For example, the cost of cutting down trees will vary with the species of tree. It takes a longer time to cut down and saw up an oak tree than a pine tree, because of its harder wood. It will cost more to get logs and other products from a woodlot that is on a steep slope or that is rocky or covered with brush than from a woodlot on level land, clean and open. The distance necessary to haul the products to the mill or market will vary with different woodlots. The skill of the workmen in the woods, the cost of labor, the climate and the size of the operation are other factors that must be considered.

The following data, taken from the second Report of the State Forester of Kentucky, will give some idea of the cost of different operations in the woods:

COST OF LOGGING (HARDWOODS)

	Cost per 1000 Bd.ft.		
	Low.	High.	Avge.
Felling and cutting into logs	$ .75	$1.75	$1.25
Skidding (often omitted or included with hauling expenses) (including swamping)	2.00	4.00	3.00
Hauling, 1 trip a day *:			
1000 feet per load	2.00	4.00	3.00
800 feet per load	2.50	5.00	3.00
600 feet per load	3.33	6.67	5.00
400 feet per load	5.00	10.00	7.50
200 feet per load	10.00	20.00	15.00
Loading		1.00	.50

MANUFACTURE AND HANDLING OF LUMBER

	Cost per 1000 Bd.ft.		
	Low.	High.	Avge.
Sawing and stacking	$3.00	$4.00	$3.50
Planing and finishing	1.00	2.00	1.50
Handling lumber in yard (including grading and loading	1.00	1.00	1.00
Hauling, 1 trip a day *: (1000 board-feet per load)	2.00	4.00	3.00
Loading on cars	.50	1.50	1.00

COST OF TIE PRODUCTION

(For 7×9 tie; 6×8 tie one-fifth less)

	Cost per Tie.		
	Low.	High.	Avge.
Cutting and hewing ties	$ .12	$ .18	$ .15
Cutting and sawing ties by portable mill	.15	.20	.18
Hauling, 1 trip a day *:			
10 ties	.20	.40	.30
15 ties	.15	.30	.23
20 ties	.10	.20	.15
Loading on cars	.03	.05	.04

* For more than one trip a day divide these amounts by the number of trips per day. For less than one trip a day multiply these amounts by the number of days per trip.

COST OF POLE AND PILE PRODUCTION

(35-feet length, 6 inches top)

	Cost per Pole or Pile.		
	Low.	High.	Avge.
Cutting and trimming	$ .20	$ .40	$ .30
Hauling, 1 trip a day †:			
3 poles per load	.66	1.33	1.00
4 poles per load	.50	1.00	.75
5 poles per load	.40	.80	.60
Loading on cars	.05	.15	.10

† For more than one trip divide these amounts by the number of trips.

COST OF PRODUCING MATERIAL PER CORD

	Cost per Cord.		
	Low.	High.	Avge.
Cutting (per stacked cord of 4-foot wood)	$ .75	$1.25	$1.00
Hauling, 1 trip a day	2.00	4.00	3.00

Peeled poles, ready to be hauled out of the woodlot.

CHAPTER XVI

MARKETING THE TIMBER AND THE PRODUCTS OF THE WOODLOT

THERE are several methods of selling the trees standing in the woodlot, or the stumpage, as it is called. The common method is to sell the timber by the boundary for "a lump sum" or "by the lot" or "on the stump." A sawmill man goes through the woodlot usually without the owner being aware of the fact, and estimates the timber. He records every saw-log, pole, tie, post or other product that he can cut from it. He figures carefully and closely just what profit can be made from the woodlot. He then offers the owner a certain sum for the entire tract. Knowing nothing as to the quantity of material in his woodlot and less about the value of the different kinds of woods or their use for special purposes, the offer may seem large to the owner and he accepts it. Such a sale is always in the interest of the purchaser. Lumbermen figure on at least a 25 per cent leeway. There is rarely any competitive bidding for the timber. The purchaser makes his own bargain.

A second method is to sell the stumpage according to an estimate made by some competent forester or practical woodsman who is acceptable to both buyer and seller. This is a method in common use. Where the estimator is reliable and has good judgment in regard to allowing for defects and rot in trees, and is a careful and skillful worker, this method is a fairly satisfactory one. It gives the least trouble to the woodlot owner. Where restrictions are made as to the kinds and sizes of trees to be cut, some supervision will be necessary. In this case it will be better to have some competent person go through the woodlot and mark all the trees to be removed.

The only safe and equitable way of selling timber is to have the purchaser pay for the material as it is cut and removed from the woodlot or as delivered at the mill, so much per piece or per unit amount, the method of logging and the price for each kind of material to be agreed on beforehand in writing. The products of the woodlot would be measured by the woodlot owner and the purchaser, or by a log scaler acceptable to both parties, and by the use of a certain log scale or other unit in general use. This is the method of selling timber used by the United States Forest Service. An estimate is made of the timber before the sale, to serve as a basis for making a bargain and setting a price on the timber. The material is paid for, however, according to the scale sent in by the government scaler, who determines the number of board feet in every log by a log rule and measures the number or lengths of all sticks of timber removed.

THE STUMPAGE VALUE OF TIMBER

The stumpage value of timber in a woodlot will depend on several factors which will vary for different regions and conditions.

On the Amount of Standing Timber in the Region.—The stumpage price of timber has gradually risen as the old virgin timber has been cut. In regions where much of the old timber still remains the stumpage price is low, as in the South, where the stumpage price of timber is often as low as $1 per thousand board feet. In the Lake States nearly all the virgin pine stands have been cut and stumpage values for good white pine timber run as high as $15 per thousand.

On the Demand for the Timber.—Where there is a ready market for the sale of the products of the woodlot, the stumpage price will be high. Often individual trees of good form and quality bring a very high price, as in the Ohio Valley, where owing to the scarcity of timber and the necessity of freighting logs from the Appalachian Region, trees in the farm woodlot are in great demand by wood-using industries.

On the Nearness to Market.—The length of the haul directly affects the stumpage price. If but one trip can be made to market a day standing timber will be worth less than where two trips a day can be made.

On the Species of Trees and the Condition of the Timber.—Not all trees have equal commercial value. The woods of some species have many uses while the woods of other kinds of trees are of little if any value. Where trees are adapted to a special use the stumpage price will be high. Sound timber brings a higher price than defective, diseased or insect-infested timber.

On the Amount of Timber Per Acre and the Extent of the Tract.—The more timber there is in a woodlot the better price it will bring. This is owing to the fact that the more timber that can be hauled over the same log road or that can be sawed by the same set-up of the sawmill, the less the expense in manufacturing the lumber.

On the Logging Cost.—The greater the cost of logging a tract the lower the stumpage price will be. With the same kind and character of timber the stumpage price will be less on steep rocky land than where the logging conditions are better.

On Whether all the Timber is to be Sold or Some Reserved.—Where seed trees are to be left to seed up the area or small trees below a certain diameter limit are to be left to grow to larger size, the stumpage price will be less than where all the timber is sold. This is not only due to the smaller amount of timber to be logged, but also because of the care that must be taken in felling trees and in logging not to injure the trees left.

On the Length of Time Allowed to Cut the Timber.—The shorter the time allowed, the lower the stumpage price will be. Purchasers of woodlots will often buy the trees and hold them indefinitely without cutting, waiting for a rise in the stumpage price. More has often been made from a rise in stumpage price than from the profit in manufacturing the material into lumber.

In every region and locality the stumpage price is determined by these different factors and is a more or less fixed amount for the same kind of timber. Stumpage prices do not fluctuate rapidly. In spite of changes that may have occurred, there has been a gradual rise in the value of standing timber since the early days, and the price will undoubtedly continue to rise as timber becomes scarcer. In many regions the valuable timber now lies in the farm woodlot and its value is bound to increase in the future. No woodlot owner should sell his timber until he knows that he is getting all that the timber is worth.

HOW TO FIND THE STUMPAGE VALUE OF TIMBER

The stumpage value is found by subtracting from the selling price of the manufactured product, the cost of logging and the cost of manufacturing, and also the depreciation on the sawmill, the logging outfit and equipment, and the profit. The profit is often figured at 25 per cent of the logging and manufacturing costs. This can be expressed in a formula:

Stumpage price = selling price — (cost of logging + cost of manufacturing + depreciation) — profit.

On a portable sawmill operation in the northern Appalachian Region the stumpage value was figured as follows:

Cordwood, poles, and ties from farmers' woodlots ready to be shipped to the market.

Measuring stacked cordwood.

	Bd. Ft. per M.
Cost of logging, including hauling to the mill..	$ 4.00
Cost of manufacturing, including handling, hauling to cars and loading................	8.00
Profit 25 per cent of costs....................	3.00
Total expenses	$15.00
Selling price for mill run....................	$20.00
Total expenses	15.00
Stumpage price	$ 5.00

MARKETING THE PRODUCTS OF THE WOODLOT

The problem of marketing the products from the woodlot is similar in many respects to that of marketing agricultural products from the farm. If the markets are good it is a simple matter, if not, the same difficulties are met. The products are not ordinarily sufficient in amount to attract the attention of the large buyers, and owing to the character of the material much of it being of comparatively little value in proportion to its weight, it cannot be shipped far. The market available to the woodlot owner is usually a local one, and the demand is soon satisfied. In addition to this, as far as sawed material is concerned, is the prejudice against the product of the portable sawmill, because of the lack of uniformity in size, thickness and quality.

Sawmill men are able to overcome many of the difficulties met by the woodlot owner. Buying and selling the products of woodlots is a business with them. They make a study of the market conditions and know all the users of wood products and the wood-using industries within striking distance, and have a knowledge of the kinds and quality of material each one needs. They make a study of the prices offered for different products and manufacture the material accordingly, and can usually sell to their own advantage. Cutting many tracts in the same region they can store the material until they have enough to make a sale and a shipment often to a distance. They can sort the sawed material and accumulate sufficient amounts in different grades to sell for special purposes and so secure a much greater profit than if they sold the mill run or the lumber as it leaves the tail of the mill, all grades and kinds mixed in together, as must usually be done by the woodlot owner. Being skilled in the sawing of lumber and the care and adjustment of machinery, they can produce as uniform and as well manufactured a product as the larger mills.

Woodlot owners in the same region by combining their efforts can overcome many difficulties in regard to manufacturing and selling their products. A sawmill can be purchased in common or a reliable sawmill man can be induced to locate in the region. The cutting of the trees and the hauling of the material to the mill can be carried on under a competent "woods boss," the woodlot owners with their help forming the different crews. The sawmill man should be an expert in studying and searching out the best markets for the different products.

Woodlot owners should not be in a hurry to harvest and market their timber. They should know before selling that they are getting the best price possible for all the different kinds of material in the woodlot. Trees are not perishable material like many farm products and they increase in value by their annual growth. In addition to this is the constant rise in the value of the stumpage. Time should be taken to make a complete study of the various markets, the requirements of each, and the prices paid for the different products. Trees should not be sold when the market is down, but should be held for more favorable conditions.

The same is true of the products of the woodlot. They should be marketed only when a fair price can be obtained.

The sawed products can be graded and stacked in the yard and held for a good market. Most sawmill operators fail not because of inability to manufacture good material, but from the failure to secure good markets and a profit on the material sold. A lumber operation should be handled only as a business. The plans and contract should be drawn up and submitted for consideration to some competent authority. Information in regard to the selling of timber and products from the woodlot can be obtained free of charge from the State Forester or from the Forestry Department of the Agricultural College. All the available information in regard to the users of wood products should be obtained from these sources and also from the United States Forest Service, Washington, D. C.

THE CONTRACT OR AGREEMENT

Before entering into an agreement for the sale of timber the woodlot owner should discuss with the purchaser all the important features of the work in the woods, the methods to be employed, the disposal of the products, and the financial side of the operation. Their agreement should be put in the form of a contract to be signed by both parties. Even in a small sale much trouble and often financial loss can be avoided by putting the terms in writing.

A sample form of timber sale contract has been drawn up by the United States Forest Service, Washington, D. C., for the use of woodlot owners. This can be secured free of charge. It covers all the different methods of selling timber from the woodlot and puts into proper form the different points that should be included in a contract for the proper carrying out of the operation and the protection of the future of the woodlot.

CHAPTER XVII

THE CONTENTS OF LOGS AND TREES

Units for Measuring Wood.—*The Board Foot.*—The board foot is a board 1 foot square and 1 inch thick. It is the common unit for buying and selling lumber in Canada and the United States. It is also used for measuring the amount of lumber that can be sawed from logs and trees. As applied to lumber and squared timber it is a measure of the solid contents. If a board contains 60 board feet it means that the actual contents of the board is 60 board feet. If a log is said to contain 60 board feet it means that 60 board feet can be sawed out of the log. The contents in board feet of lumber and squared timbers is found by multiplying the length in feet by the width in feet, by the thickness in inches. In measuring lumber no allowance is made for defects, but the material is classified into grades according to the number, size and character of the defects. With inch boards the contents in board measure is the same as the number of square feet of surface, and is often spoken of as surface measure. The number of board feet that can be sawed from logs is shown in log tables.

Cord Measure.—Firewood, pulp wood and other material sawed into short lengths or bolts, as excelsior wood, spool wood, novelty wood, etc., are usually measured by the cord. Cordwood is usually cut into 4-foot lengths. A cord of such wood will be a stack 4 feet high, 4 feet wide and 8 feet long. It will contain 128 cubic feet. A short cord is one in which the pieces are cut shorter than 4 feet. Stove wood is sometimes cut 14 to 18 inches long.

The Relation Between Board Feet and Cordwood.—No definite relation exists between the cord and the number of board feet it contains. The number will vary with the character of the wood, the method of piling used and other factors. In general for ordinary sized wood a cord is said to contain about 550 board feet of wood.

The Cubic Foot.—The cubic foot has never come into general use in this country for measuring wood, although it is the best unit for measuring the volume of logs, for it gives the entire contents of the logs. It is being used at the present time for measuring pencil wood, wagon stock and other valuable hardwoods, especially imported woods. Pulpwood is frequently measured by the cubic foot. In many countries of Europe and in the Philippine Islands the cubic meter is the standard measure for finding the contents of logs and timber. The cubic contents of trees is used by foresters in estimating the contents of a stand of timber for cordwood and in scientific work.

Linear Measurement.—Many products of the woodlot are measured by the length in feet instead of by the contents. Poles, mine props and piles are usually measured and paid for by the piece or by linear measurement. Ties, fence posts and many other products are often bought and sold by the piece according to size.

Finding the Volume of Logs.—*Cubic Measure.*—There are two methods in common use for finding the contents of logs.

In one method the log is considered as a cylinder with a base corresponding to the diameter at the middle of the log. The volume is found by measuring the diameter at the middle of the log accurately and finding the area of the base in square feet. This, multiplied by the length of the log in feet, will give the cubic contents of the log. (Basal area tables corresponding to different diameters will be found in the appendix.)

Expressed as a formula:

$$V = B_{\frac{1}{2}} \times L$$

in which V is the volume of the log in cubic feet. $B_{\frac{1}{2}}$ is the area of the middle cross-section of the log in square feet. L is the length of the log in feet.

A more accurate method, but requiring one more measurement, is to average the basal areas of the small and large ends of the log and then multiply by the length of the log. The average diameters at the top and bottom of the log are measured accurately and the corresponding basal areas in square feet are found from the basal area tables. These are added together and divided by two to get the average basal area. The result is multiplied by the length of the log in feet to get the cubic contents of the log.

Expressed as a formula:

$$V = \frac{B+b}{2} \times L$$

in which V is the volume of the log in cubic feet. B and b are the areas in square feet corresponding to the diameters of the two ends of the log. L is the length of the log in feet.

Board Measure.—The number of board feet that can be sawed from logs of different diameters and lengths is shown in tables called log scales or log rules. A log table or rule is a tabulated statement of the contents in board feet of logs of different diameters and lengths. Knowing the diameter of the small end of a log inside the bark and the length of the log, the number of board feet contained in the log can be found from a log rule. There are over forty such rules in common use, often differing widely in the number of board feet given for logs of the same size. The amount of lumber that can be sawed from a log of a certain size is not definite.

It will vary with many factors, such as the thickness of the saw, the efficiency of the machinery, the skill of the sawyer, the thickness of the boards sawed, the defects in the logs and others. This will account in some measure for the many rules in existence.

Log rules are constructed in several ways, some by mathematical formulæ some by diagrams of the small ends of logs, some are based on actual experience at the mill in sawing boards from logs, and others by correcting existing log rules to fit local conditions.

The following log rules are in most general use in the eastern part of the country:

The Scribner Rule.—The Scribner rule was constructed by the use of diagrams. It is the oldest rule in general use. Circles were drawn the size of the small ends of logs of different diameters and in these circles the inch boards that could be sawed from them after allowing for waste in sawing were mapped out. From these the number of board feet that could be sawed from logs of different lengths were found. The results for logs of different diameters and lengths were then arranged in a form of a log table.

A modification of this rule, called the Scribner Decimal Rule, has been made by dropping the unit figures and rounding the values to the nearest tens. This is the log rule adopted by the United States Forest Service for timber sales on the National Forests. The Scribner Rule is said to give very fair results for logs below 28 inches in diameter, but for logs above the results are too low.

The Doyle Rule.—The Doyle Rule is based on the following mathematical formula: Deduct 4 inches from the diameter of the log inside the bark to allow for slab, square one-quarter of the remainder and multiply this result by the length of the log in feet. This rule gives very low results for small logs, and above 36 inches in diameter the values given are too large. This is because the width of the slab removed by the formula is the same regardless of the size of the log. Too much allowance is made for small logs and not enough for large logs. Yet this rule is in more general use throughout the country than any other. The mill cut will often overrun the values given by Doyle Rule by 25 per cent.

The Doyle-Scribner Rule.—A combination of the Doyle and Scribner Rules, known as the Doyle-Scribner Rule, is in common use for measuring hardwoods. In this rule the figures of the Doyle Rule are used for small logs under 28 inches in diameter, and above 28 inches the Scribner Rule is used. Since the Doyle Rule gives low results for small logs and the Scribner Rule gives low results for large logs, this rule is evidently unfair to the one selling timber.

The International Log Rule.—Several attempts have been made to construct a standard log rule that will state with fairness to both buyer and seller the contents of logs. Such a log rule is the International Rule, which is based on a mathematical formula checked by actual experience in sawing out boards from logs. Several log rules will be found in the Appendix.

Considerable confusion often arises in measuring the contents of logs through the use of different log rules by the buyer and seller. The log rule to be used should be specified in the contract to be drawn up for the sale of the timber.

Scaling Logs.—The measuring of logs to find the contents in board feet by means of a log rule is called log scaling. For convenience in the woods the log rule is printed on a wooden stick about 4 feet long, called a scale stick. By means of the scale stick placed across the small end of a log inside the bark the width rounded to the nearest inch is noted. Opposite the inch mark will be read in line with the proper length of log the number of board feet or the scale of the log.

A log rule is constructed to give the contents of perfect logs without crook or other defects. If defects occur in logs the scale as given in the log rule must be discounted to allow

Grades in southern yellow pine determined by the number, size, character and position of knots and other defects.

for them, otherwise the values will be too high. What is wanted in scaling logs is the number of board feet that can actually be sawed out of the logs. No rule can be given for deducting for defect in logs. It is a matter of judgment. If it is believed only half the log will cut sound wood, only half the scale should be recorded. Much can be learned as to allowing for defects by watching logs sawed at the mill, and watching the number of board feet thrown out because of a certain sized defect. Crook is often allowed for by reducing the diameter of the log by enough to square up the log.

Rule of Thumb for Finding the Contents of Logs in Board Feet.—The following rule of thumb is a very simple method of determining without the aid of a log rule the contents of logs in board feet. For logs that are 16 feet in length, subtract 4 inches from the diameter of the small end inside the bark and square the remainder. It gives the contents in board feet of logs according to the Doyle Rule. For example, if a 16-foot log is 14 inches in diameter the rule will work as follows: Four inches subtracted from 14 inches leaves 10 inches, and 10 squared gives 100 the number of board feet in the log. If a log is 12 feet long, or any length other than 16 feet, apply the rule as though the log were 16 feet long, then take the part of the scale that the length of the log is of 16. If the log is 12 feet long three-quarters of the scale of a 16-foot log of that diameter is used. If a log is 8 feet long one-half the scale is taken.

Finding the Volume of a Tree.—*The Cubic Contents.*—The usual form of a tree is a paraboloid or a cone with slightly bulging sides. The only way to find the actual solid contents of an entire tree would be by displacement of water, cutting the tree into parts and placing them in a tub of water and measuring the amount of water in cubic feet displaced or that runs out. It would be almost impossible to measure the limbs and twigs accurately in any other way. Such a high degree of accuracy is not necessary in measuring a standing tree for purposes of estimating the contents. When the entire tree is considered as a paraboloid it means that if the twigs and limbs were all squeezed up together they would carry up the solid trunk of the tree with a gradual taper to the tip.

If the solid contents of a cone is found by multiplying the area of the base by one-third the height, then the volume of the tree, which is a little larger than the cone, will be found by multiplying the area of the base in square feet measured at breast height by one-half the total height of the tree.

Expressed as a formula:

$$V=\frac{B\times H}{2}$$

in which B is the area of the base of the tree in square feet corresponding to the diameter measured at breast height, and H is the total height of the tree.

The Volume of a Standing Tree in Board Feet.—The volume of a standing tree in board feet can be found by dividing the merchantable trunk of the tree into logs by the eye, estimating the top diameters inside the bark of each log and finding the board feet from a log table or by applying the rule of thumb for board feet in logs. The result of the different logs added together will give the estimate in board feet in the tree.

A rule of thumb well known among timber estimators for finding the volume of the merchantable length of trees in board feet is to subtract 60 from the square of the estimated diameter inside the bark at the middle of the merchantable length of the tree and multiply the remainder by .8. The result will be the number of board feet in the average 16-foot log in the tree. This number is then multiplied by the number of 16-foot logs in the tree, which will give the board-foot contents in the tree.

Volume Tables.—Foresters make use of volume tables in estimating large tracts of timber. These give at once the contents of a tree either in board feet or in cubic feet or other unit when the height and diameter of the tree are known. The diameter is always measured at breast height and is the average of the largest and smallest diameters at that point if the tree is not round. The height may either be the total height of the tree or the merchantable length of the trunk. Each volume table will specify which to use. The simplest form of volume table is based on diameter alone. Volume tables give average results, for the figures in volume tables are found by averaging together the volumes of a large number of trees of the same height and diameter that have been cut down and measured. They are intended for estimating only a large number of trees. They are not to be used for measuring the contents of an individual tree. Volume tables give very accurate results when used for estimating large stands of timber.

Measuring the Diameters of Trees.—The diameters of trees are always measured at breast height or four and a half feet above the ground. This point will be well above the flare or swell usually found at the base of a tree. The trunks of most trees are not perfectly round in cross-section, so that two measurements are always taken at right angles to each other and the average diameter used.

The instrument used for measuring diameters of trees is a wooden calipers. These can be purchased for a few dollars. For estimating a small woodlot two carpenter's squares placed end for end and the arms against the tree will measure the diameter with great accuracy. Wooden calipers can easily be made.

Measuring the Heights of Trees.—The heights of trees are measured by instruments called hypsometers. These are of several forms and are known by different names. A simple one is the clinometer, which when sighted at the tip of a tree from a distance of 100 feet will read directly the height of the tree.

There are several methods for finding the heights of trees without instruments. One of the simplest methods is to measure the length of the shadow cast by the tree and the length of the shadow cast by a 10-foot pole set perpendicularly in the ground. When the length of the shadow cast by the tree

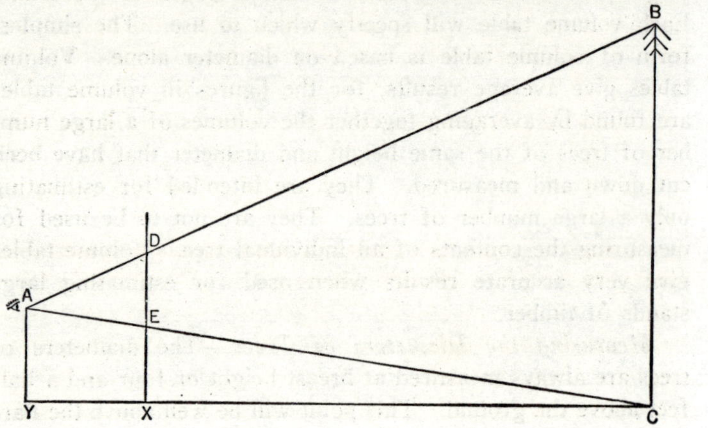

is multiplied by the length of the pole and the product divided by the length of the shadow cast by the pole the result will be the height of the tree. The heights of trees can be found in this way only when they stand in the open.

A staff about 4 feet long, one end held to the eye and grasped at arm's length and then held vertically in front of the eye, can be used in measuring the height of trees. A position is found out from the base of the tree where the top of the tree will be seen just over the top of the pole and the base of the tree over the hand. The distance the observer is then away from the tree will be the height of the tree.

The height of a tree can also be found by the use of two staffs. The observer moves away from the tree about 50 to 75 feet until he can see the top and base of the tree. A staff 6 to 8 feet high is then set upright at this point (x in diagram). A few feet farther away from the tree a second staff about 4 feet in height is set at (y). The observer now looks from the top of the staff (y) at A across the first upright staff (x) to the top of the tree (B) and an assistant marks on the staff (x) the point (D) where this line crosses. The observer now looks from (A) to the bottom of the tree and the assistant marks on the staff (x) the point (E) where this line crosses.

The distance AC or from the eye of the observer to the foot of the tree multiplied by DE the length marked on the staff (x) and this product divided by AE, the distance from the eye of the observer to the staff (x), will give the height of the tree.

Height of tree: $$BC = \frac{AC \times DE}{AE}$$

(For AC the distance yc on the ground can be measured, and for AE the distance yx on the ground can be measured.)

CHAPTER XVIII

ESTIMATING THE TREES IN THE WOODLOT FOR BOARD FEET AND FOR CORDWOOD

For many years it was thought impossible to learn how to estimate the contents of standing trees without years of practical experience in the woods. The early estimators were men who had first worked in timber helping to cut down trees and saw them into logs. Through years of observation of tracts of timber before they were cut and a subsequent knowledge of the amount of lumber each tract sawed out at the mill, their judgment became trained so that they were able to guess with a fair degree of accuracy as to the contents of the timber standing on a particular tract. These men were sometimes called timber lookers, for about all they did was to pass through the timber and look it over and then to state the amount of lumber or cordwood that it would produce. They were often also called timber cruisers, for they made their way about through the timber by means of a compass, as a ship is cruised through the water. Some of these early timber estimators were held in high repute for their ability to state with considerable accuracy the contents of forests. But usually the results were far from being right. These early methods were good enough when standing timber had but little value and was often sold by the acre. When timber became scarce and the stumpage value had risen so that even individual trees in the woods had a value, the old methods of timber estimating were no longer acceptable. In purchasing timber men no longer cared to risk their money on a guess by a timber looker. Methods of estimating having some scientific basis were then devised. These give with assurance when carefully carried out the contents of stands of timber usually within at least 5 per cent to 10 per cent of the actual amount. Most of these methods can be used without previous experience in the woods.

In estimating timber no effort is made to measure the trees exactly. This could be done only at great expense by cutting the trees down or climbing into them. An estimate, not the exact contents of the trees is desired. A timber estimate is made usually to give a basis for determining the stumpage price or the value of the standing timber and for figuring on the best method of cutting the trees and getting the logs out of the woods. Settlement for the logs cut is usually made by measuring the actual amount of material removed from the woodlot.

There are many methods of timber estimating in use at the present time. Each timber estimator chooses the method best suited to the timber, the object of the estimate, and the time and money at his disposal. These methods will be found described in books on Forest Mensuration and in books on General Forestry, a list of which is given in the Appendix. From these many methods a few have been selected which are simple in theory and easy to apply, and which can easily be carried out by the farmer in the woodlot. Although these methods are simple and easily learned they are among the most accurate known. Where carried out skillfully and carefully they will give with considerable certainty the contents of the timber.

Value of Timber Estimating to the Farmer.—Every woodlot owner should be able to estimate the contents of the trees in his woodlot. He should know how much lumber the trees will produce when sawed and also how many cords of wood or how many poles, ties and other products. Without such knowledge in selling his product he usually is dependent on guess work or on the estimate made by the purchaser. A knowledge of the forest capital contained in his woodlot will enable him to figure the amount that he can cut safely each year from the woodlot.

ESTIMATING A SMALL WOODLOT FOR LUMBER IN BOARD FEET

In estimating a woodlot for board feet only that portion of each tree is considered that will make saw logs. The tops and limbs can, of course, be used for cordwood or other purposes. What is desired is a knowledge of how many board feet of boards can be sawed from the sawlogs contained in the trees. The number of ties, posts and other products, in addition to the sawlogs, can of course be determined at the same time the estimate is made for board feet if desired.

In estimating a small woodlot for board feet each tree is considered by itself, and when it has been estimated it should be marked or chalked in some way, so that it will not be measured again. The standing tree to be estimated is divided into logs by the eye into even foot lengths, and into 16-foot lengths as far as possible. The top of the first 16-foot log, or butt log, can easily be found by using a 10-foot pole. When held by an assistant 6 feet above the ground, or the top of the stump, the top of the pole will indicate where the first 16-foot log will end. The lengths of the logs above can be estimated by the eye with considerable accuracy when standing about 50 to 75 feet from the tree. In this way the number of logs that can be cut from each tree and their lengths can be found. The diameter of the small end of each log inside the bark must be known as well as the length. The diameter of the butt lot can be measured accurately by nailing a cross-piece marked plainly in inches on the upper end of the 10-foot pole. When held by the assistant so that the cross-piece is 16 feet above the stump, the observer can readily count the number of inches in diameter of the top of the log. The measurement must be inside the bark. Very few trees except old trees have a bark 16 feet above the stump that is more than a half inch thick. For most trees an inch should be subtracted from the diameter outside the bark to allow for the thickness of the bark. The diameters inside the bark at the tops of the logs above the butt log must be estimated by the eye. The upper logs are much smaller usually than the butt log. The first log or butt log in a tree is the important log in estimating and will be found to contain a much larger amount of board feet than any log above. If the diameter of .the butt log is measured accurately the diameters of the upper logs can safely be estimated by the eye. A knowledge of the diameters of upper logs as compared with butt logs can be obained by measuring trees that have been cut or blown down by the wind.

When the length and top diameter of a log is known its contents in board feet can be found from a log table or by the simple rule of thumb already described: for logs that are 16 feet in length, if 4 inches are subtracted from the top diameter and the result squared it will give the number of board feet in the log according to the Doyle Rule, and for logs of other than 16-foot lengths they are figured as though 16 feet long and the part of the scale taken that the length of the log is of 16 feet.

In estimating the contents of each log in board feet in this way it should be noted whether the logs are perfect or not. Log rules and the rule of thumb mentioned give results for perfect logs. Most logs have imperfections and allowance must be made for them. No rules can be laid down for doing this, but any one with good judgment can make a fairly accurate deduction.

In this way the board feet should be found for each log in the trees in the woodlot. The total board feet will be the estimate for the tract. In tallying the board feet for the logs those that are of the same species of trees can be kept together so that the final result will give an estimate for each kind of wood.

The following form can be used for recording the data for each log estimated. A separate form can be made out for each species, or a column can be added to indicate the species:

Species.	Diameter of log.	Length of log.	Condition of log. Defects, crook, etc. Amount to deduct.	Board-feet in log after allowing for defects.

At the same time the logs are being recorded in each tree, the estimator, if it is desired, can record the number and sizes of ties, or fence posts or other products that are contained in the trees, and also the number of poles and their lengths. A complete estimate should give the amount and kinds of all material that can be cut from the woodlot.

ESTIMATING A LARGE WOODLOT FOR BOARD FEET

To estimate each tree in a large woodlot in this way would require a great deal of time and labor. It is possible to pick out several sample acres that are representative of the entire woodlot if the trees are fairly uniform, and find the average board feet per acre. Then by multiplying by the number of acres in the woodlot an estimate for the entire woodlot will be obtained. (A simple method of finding the number of acres in a woodlot is described in the Appendix.) Great care and judgment must be used in selecting the sample acres to be estimated. They should neither be located in the best portion of the timber nor in the poorest. The sample acres should be average acres. Not less than 4 or 5 acre plots should be estimated for a 20-acre woodlot to secure a good result. The number of board feet on all the sample plots estimated added together and then divided by the number of sample acres measured will give the average number of board feet per acre. Square acre plots can be laid out 208 feet on a side.

Smaller plots can be used where the timber is very uniform. A square half-acre plot measures 147 feet on a side and a quarter-acre plot 104 feet on a side.

Circular plots can also be used. They are laid out more quickly than square plots. A circular plot covering a quarter-acre in extent will have a radius of 59 feet. This will be the most practical size of plot to use. Choosing a certain tree for the center of the circular plot or setting a stake, the radius distance is measured from the center out in several directions. Small pieces of paper fastened in the bark of trees will help to indicate the boundary line. Eight or ten such measurements made from the center and well distributed about the circle will mark the line sufficiently for the estimate. Sometimes but a single sample plot is used for estimating an ordinary woodlot, but usually the estimates on several such sample plots are averaged together. These plots can be laid off here and there in timber that appears to represent the average of the stand, or they can be laid out at regular intervals throughout the woodlot, so that when averaged together they will give an average of the entire woodlot.

The process of estimating the trees on these sample acres can be the same as for the method described for the small woodlot. Each tree on a sample acre is estimated by itself, divided into logs, and the number of board feet in each log found from a log table or by the rule of thumb. Care must be taken as before in making allowance for defects in the logs or the estimate will be too high.

Where a quick estimate is wanted the estimator can, after sizing up the trees on the plot, pick out a tree by eye that he believes to be the average sized tree of all the merchantable trees on the plot. The contents of this tree can easily be figured, and when multiplied by the number of trees on the plot the product will be the estimate for the plot. The result will be more accurate if instead of one average tree several are selected, and the contents of the average tree found by averaging together the contents of all of them. The accuracy of the estimate will, of course, depend on the skill used in picking out the average trees.

With very large tracts of timber volume tables are used instead of log tables. These give the contents of trees in board feet when their heights and diameters are known. All the trees on the tract can be estimated or only those on sample acre plots or on measured strips run in parallel directions across the tract. Volume tables for several species will be found in the Appendix. Many others are given in the Woodmen's Handbook, Bulletin 36, U. S. Forest Service, and in books relating to general forestry.

ESTIMATING A SMALL WOODLOT FOR CORDWOOD

In estimating a small woodlot for cordwood each tree is estimated by itself, as in the case with board feet. When a tree has been estimated it should be marked in some way so that it will not be considered a second time. If all the trees are marked on the same side it will be easier to keep track of them.

To obtain the cords of wood in a woodlot the cubic feet contained in all the trees are first found. The total cubic feet in the trees divided by 90, the average number of cubic feet of solid wood in a cord of wood, will give the estimate of the number of cords of wood in the woodlot.

The number of cubic feet in a tree including the larger branches is found, as already explained, by considering the entire tree as a cone with slightly bulging sides and finding the volume in cubic feet by multiplying the basal area in square feet corresponding to the diameter of the tree measured at breast height by one-half the total height of the tree in feet.

The diameters are measured accurately to tenths of inches outside the bark, for the whole tree is used for cordwood. If a tree is not perfectly round the average of the largest and

Forestry students estimating the contents of the trees in a very dense stand of pine.

smallest diameters is taken. The corresponding basal areas in square feet are found from the basal area table. The total height of each tree is found by the use of a hypsometer or by one of the other methods described in a previous chapter.

The data obtained for the different trees can be recorded in a table as follows:

Diameter of tree at breast-height in inches and tenths.	Total height of tree in feet.	Basal area in square feet from basal area table.	Number of cubic feet in tree.

A different form can be made out for each species of tree if it is desired to keep the species separate or another column can be added to indicate this. The total cubic feet in all the trees added together and this result divided by 90 will give the estimate for the woodlot.

ESTIMATING A LARGE WOODLOT FOR CORDWOOD

In a large woodlot instead of estimating each tree in the woodlot, which would be a very long operation, sample acre plots as described for estimating for board feet can be laid off in the woodlot and the average number of cords per acre on the woodlot obtained. This multiplied by the number of acres in the woodlot will give the estimate for the entire tract. Square sample acre plots or circular plots can be laid off in the woodlot located in timber that will be an average of the timber in the woodlot, and each plot estimated by itself and the average for the woodlot obtained.

Volume tables can also be used in estimating a large woodlot for cordwood. Knowing the diameters and heights of trees the contents can be found from prepared volume tables. These give the contents of trees in cubic feet and sometimes in cords of wood.

CHAPTER XIX

THE DIFFERENT PROPERTIES OF WOODS

As we have already learned, wood is not simple in its structure, but is made up of many very small cells. The cells differ in different woods and even in the same wood there may be several kinds of cells each serving a different purpose. The different kinds and arrangements of cells give rise to the different properties in woods. The reason why one wood is soft and another hard, or why one wood is strong and another weak is because each has a different structure. Often in the same kind of tree the wood will vary in its properties when grown in different situations and when grown at different rates. Even in the same tree the properties of the wood will vary in different parts of the stem. The heartwood will have different properties from the sapwood and the wood of the butt log from the wood of the top logs, the wood from the quickly grown sapling of the abandoned field from that of the slowly grown forest tree. Even the manner in which the wood was sawed from the log and the method of drying and seasoning will influence the behavior and quality of the wood. However, these differences in the same kind of wood are not very pronounced.

In the selection of a wood for a particular purpose its properties must be known. Woods often bring a high price because of their adaptability for some special use and for which no other wood will serve the purpose as well. There are no trees that produce worthless wood. Even some of the so-called weed trees in the woodlot, such as dogwood and ironwood, have special uses for which they are of value.

For the ordinary uses of woods it is seldom that fine distinctions are made. There are many woods that can be used for construction purposes, such as pine, spruce, hemlock, fir and cottonwood; nearly any hardwood will make strong and serviceable furniture; fence posts are used of any of our more durable woods. Many wood-using industries, on the other hand, demand woods that have special properties that fit them for their particular purpose, such as for pencils, musical instruments, aeroplanes, shuttles, etc. In this case, where definite qualities are demanded, a special inspection of each stick must be made.

It is not on any one quality that the value of a wood for use depends, but often on a combination of many qualities. For the use of the carpenter and builder a wood that is soft to work and to nail and fairly stiff and that is abundant and cheap is desired, a combination found in the coniferous woods. The furniture maker demands a wood that is strong, hard and tough, that will not dent easily, that will take a high polish, often of good color and grain, and that will not warp and twist when once in place. For the carriage builder a wood must be strong, hard and elastic. It must have a long fiber to give toughness and strength, and for spokes must be evenly and easily split. Ability to resist shocks and to hold spikes are demanded for woods for railroad ties, and a durable, light, soft wood for poles. Often the color of wood or its odor will give it a special value, provided it has the other qualities requisite as for some kinds of furniture, interior trim and novelties. The quality of responding to and enforcing a musical note gives some kinds of wood a value for musical instruments.

It is only within recent years that a systematic study has been made of the properties of different woods and their fitness for special purposes. The U. S. Forest Service has established a Woods Products Laboratory at Madison, Wisconsin, where woods are tested for their properties and the value of woods for different purposes determined. Formerly the knowledge had of woods came from their use alone and often erroneous ideas in regard to woods were held, as for example, the idea that the heartwood of hickory was inferior to the sapwood for use for making wagon spokes. Many woods have been found to be adapted for different uses that were formerly considered worthless, and many new uses have been found for our common woods.

THE DIFFERENT PROPERTIES OF WOODS

Weight, hardness, strength, durability, elasticity, toughness, heat production, color, odor and taste are some of the properties of woods that make them valuable for use.

Weight.—The weight of different woods varies greatly. Cork oak of the Southern swamps is lighter than cork, weighing not more than one-fifth as much as water. Ironwood weighs a third more than water. The wood substance itself is heavier than water, weighing one and one-half times as much. In the form of wood, however, because of the large amount of air enclosed in the cells, the weight of a piece of wood is usually much less than the weight of the same volume of water. Most of our commercial woods when dry weigh less than water, weighing from one-half to three-fourths as much. The weight of wood depends on the wood substance itself and the amount of water contained. Only a few of our woods will float when first cut. Most coniferous woods will float when green and a few of the lightest hardwoods, like cottonwood. This often enables them to be floated or driven down streams to the sawmill. All woods weigh much lighter when dry and seasoned than when green and full of water. Some woods lose over half their weight in drying.

Woods show a great variation in density, even in the same species of tree and also from the same tree. This variation depends on the rate of growth, quality of the locality in which the tree was grown, age of the tree, the part of the tree from which the wood was cut, and the amount of resinous material in the cell.

WEIGHTS

TABLE SHOWING RELATIVE WEIGHTS OF COMMON WOODS

Very heavy woods.	Heavy Woods.	Medium Weight.	Light Woods.	Very Light Woods.
Hickory	Ash	Western pine	Red pine	White pine
Oak	Elm	Pitch pine	Cypress	Spruce
Persimmon	Cherry	Sweet gum	Hemlock	Fir
Osage orange	Birch	Soft maple	Chestnut	Cedar
Black locust	Beech	Sycamore	Butternut	Poplar
Blue beech	Walnut	Sassafras	Tulip	Willow
	Sour gum		Catalpa	
	Honey locust		Buckeye	
	Tamarack			

Hardness.—Hardness is resistance offered by wood to having the fibers pushed apart or indented. It depends principally on the density of the wood. Woods having cells that have thick walls as in hickory are harder than those whose cell walls are thin as in pine. Hardness is a property essential in furniture, so that it will not become indented easily by blows. A floor should be hard as well as strong for the same

TABLE SHOWING HARDNESS OF THE COMMON WOODS

Very Hard Woods.	Medium Hard Woods.	Soft Woods.	Very Soft Woods.
Hickory	Ash	Chestnut	White pine
Dogwood	Oak	Tulip poplar	Willow
Ironwood	Elm	Sweet gum	Cedar
Sugar Maple	Beech	Southern and Western yellow pine	Redwood
Sycamore	Cherry	Larch	
Black locust	Birch	Basswood	
Persimmon	Blue beech	Hemlock	
Osage orange	Honey locust	Cottonwood	
	Soft maple	Spruce	
	Holly	Fir	

reason. Strong woods are usually hard woods, and soft woods do not have much strength. Heavy wood is harder than light wood, especially is this true in the same kind of wood. The wood of the butt log is harder than that of the upper logs, the dark summer wood of the annual ring is harder than the light colored spring wood.

Durability.—Different kinds of woods vary in their durability or the resistance offered to decay when subjected to moist conditions. The question of the durability of woods is of importance to farmers because of the necessity of using woods in contact with the soil, as for posts, poles, stakes, sills, for silos and other purposes. Some woods are very durable while other woods soon decay when kept moist. Just what makes one wood more durable than another is not known. It is believed that during the change from sapwood to heartwood certain tannins, gums and resins are deposited in the walls of the cells along with coloring material which tend to preserve the wood. It may be that more of these substances are deposited in some wood than in others, or that these substances vary in kind. As a general rule dark colored woods are the more durable, as black walnut, mahogany, red wood, red cedar and locust. Light colored woods are usually perishable, like cottonwood, willow, birch and sycamore. Outside of this fact there seems to be but little relation between the physical properties of different woods and their durability. For example, it might be imagined that heavy woods would be more durable than light woods, but hickory and hard maple, two of our heaviest hardwoods, are not durable, while catalpa and cedar, two of our lightest woods in weight, are very durable. Neither are all soft woods perishable, for cedar, redwood and catalpa are all soft woods. With the same kind of wood, however, the heavier and harder the wood the more durable it will be. That is, a heavy piece of pine is more durable than a light piece.

The time of year when a tree is cut is often said to have

a marked influence on the durability of the wood. Many claim if posts are cut in winter, when the "sap is down," they will be more durable than if cut in summer. In the winter time the wood of trees contains large amounts of starches and sugars and other food materials stored in the wood. So a post would naturally be more durable if cut in summer than in winter. There is no meaning to the expression when the "sap is down," for the sap never moves down the wood of the tree as we have learned, and there is actually more moisture in a tree in winter than in summer, because of the less transpiration taking place from the tree. If cut in winter posts dry out slowly and do not season check so deeply as when cut in summer. This may have some effect on the durability. It is claimed by many that well seasoned posts will be more durable than posts set green. Some claim, also, if a post is set upside down it will last longer than if set with the large end in the ground. In all probability these will have little effect on the durability of the wood. Posts cut early in the spring are easier to peel than if cut at any other time.

The following table shows the relative durability of some of the common woods:

Very Durable.	Durable.	Perishable.	Very Perishable.
Black locust	White oak	White elm	Cottonwood
Red cedar	Cherry	Beech	Willow
Cypress	Larch	Hickory	Aspen
Redwood	Honey locust	Hard maple	Black gum
Catalpa	Black walnut	Red gum	Jack pine
Sassafras		White ash	Box-elder
Osage orange		Hemlock	Soft maple
Chestnut		Spruce	
		Yellow birch	
		Soft maple	
		Sycamore	
		Basswood	

Strength.—Strength in wood varies with the density or weight. Light woods lack the strength of heavy woods.

Within the same kind of wood the strength will vary with the density. With woods that form large pores in the spring, like oak, chestnut and hickory, the wood from fast-growing trees is stronger than from slow-growing trees. This is because of the greater percentage of hard, strong summer wood. The amount of spring wood will remain the same in these woods whether the wood grows fast or slowly. In the fast-growing wood the increase will almost entirely be made up of the summer wood. Second growth hickory is preferred to the forest grown wood for wagon stock because of this fact. It is stronger because of the faster growth. With conifers, on the other hand, the faster the growth the weaker the wood. In a fast-growing conifer there is a greater percentage of spring wood which is much softer and weaker than the summer wood. In conifers fast growth means a large growth in the spring wood. Since summer wood is darker than spring wood, the darker the wood in the same species, leaving out of account resin content, the stronger it will be. In selecting a strong piece of pine, then, a dark-colored piece that has narrow annual rings should be chosen.

With hard woods that do not produce large spring pores like tulip, poplar, maple and birch rapidity of growth has little effect on the strength or durability of wood.

Defects in boards or timbers decrease the strength, such as knots, checks, splits, etc. Clear lumber is stronger than knotty material. With timbers the position of the defect has some influence on the strength. A timber laid so that the defect is near the top when set will be stronger than the same timber set with the defect below. This is because in a timber subject to a stress the upper half of the timber will be subject to compression, while in the lower half of the timber the fibers will have a tendency to be pulled apart, being subjected to a tension. The manner of sawing lumber from logs and the direction of the annual rings passing through the wood will also influence the strength.

Elasticity.—Elasticity is the property possessed by many woods so that when bent out of shape they will spring back to their former position. Elasticity is essential where woods are subject to strains that tend to change their shape. The rims and spokes of a buggy wheel, the thills and tongues of wagons, an ax handle, a bow, oars to a boat, all must be made of woods that have elasticity. If woods are bent too far out of shape they will not return exactly to their former positions, but will have a permanent set.

TABLE OF ELASTICITY OF THE COMMON WOODS

Woods with Considerable Elasticity.	Woods with Little Elasticity.
Yew	Cottonwood
Larch	Birch
Hickory	Maple
Osage orange	Elm
Lance wood	Alder
Ash	Walnut
Red cedar	Yellow pine
Fir	Tulip poplar
Locust	Beech
Chestnut	
Spruce	
White pine	
Oak	

Flexibility.—Flexibility is the property some woods possess by which it is possible to bend them without their breaking. Hickory is flexible, pine is brittle. It depends largely on toughness in the wood or the ability of the fibers to hold together. Young shoots are more easily bent than old wood, moist wood than dry wood. When wood that is green or that has been moistened by steaming or soaking is bent and then dried in that position, it will always tend to retain its new shape. The rims of buggy wheels, hoops, shafts, shinny sticks, the curved handles to canes and umbrellas are examples of woods that have been bent and dried in position. Lumber that has been dried flat will usually not warp. Boards are often weighted down or clamped to prevent their warping in drying.

Toughness.—A property desired in many woods that are subject to sudden shocks so that they will recover from them is toughness. The rims and spokes of a wagon are subject to many sharp blows in every mile, an ax handle is jarred with every stroke, a trunk must stand many bumps and blows in the course of a journey. A wood that shatters easily or that breaks instead of resisting a blow will not serve these purposes. A tough wood must be strong and pliable and not weak and brittle. Basswood, hickory and ash are examples of tough woods.

Fuel Value of Woods.—In general the heavier the wood the greater will be its fuel value. All woods do not burn in the same way. Chestnut, larch and hemlock crack in the fire and emit sparks, while hornbeam, birch and alder burn with a particularly quiet flame. With coniferous woods the larger the quantity of resinous material in the wood, other things being equal, the greater the fuel value. Fat pine or light wood is wood that has a large amount of resinous material. Dry woods burn much better than wet woods and send out more heat. This is because with wet woods much of the heat will go to evaporate the moisture in the wood. It is claimed that 5.2 pounds of dry wood is equivalent in the amount of heat produced to one pound of hard coal.

TABLE OF FUEL VALUE OF COMMON WOODS

Best.	Good.	Moderate	Poor.
Hickory Beech Hornbeam Locust White oak	Ash Birch Sugar maple Elm Black walnut	Chestnut Hemlock Catalpa Box-elder Butternut Soft maple	White pine Cottonwood Aspen White spruce White fir

Color.—Color adds to the value of wood for decorative purposes and is an aid in identifying woods. The color of woods varies from the deep black of the heartwood of persimmon to the almost white wood of holly. When wood is first formed from the cambium layer it has but little color and is called sapwood. Later the wood becomes darker, changing to heartwood. The change in color is due to the infiltration of various coloring materials, oils, gums, tannins, etc., in the walls of the cells of the wood, the color varying with the kinds and amounts of these materials.

Odor and Taste.—Nearly all woods have a characteristic odor and taste. In some, like red cedar, sandalwood and camphorwood the odor adds to their value. Rotting wood also has an odor that is characteristic of the species, as has also the smoke of burning woods. Many woods have a pronounced taste—willow a bitter taste, chestnut an astringent taste. The taste and odor of woods also aid in their identification.

CHAPTER XX

THE USE OF PERISHABLE WOODS FOR FENCE POSTS BY TREATING WITH CREOSOTE

In many regions the question of a future supply of fence posts is becoming a serious one. Trees yielding durable woods have been plentiful in the past, but the time has come when many of them are too scarce and too high in price to be utilized generally by farmers. White oak, formerly one of the chief post woods, is becoming too valuable for other purposes to be utilized for fence posts. The supply of chestnut is sure to be greatly depleted by the chestnut blight, if not entirely wiped out. Locust for fence posts is now hard to obtain and in some sections the price is almost prohibitive. Cedar has long since ceased to be a fence post material except locally. Its use for making pencils and for boxes and chests has raised the value of the wood too high for general use. Walnut, one of the most durable of woods, has become very scarce.

Because of this condition substitutes for the wooden posts are being sought. Many posts of iron and concrete are being made and used, but their cost is usually high. Concrete posts can be made for about 25 cents each. But even at that price they will not come into general use. A fence post to be practical must be cheap. In foreign countries where wood is too valuable and scarce to be utilized for fence posts, farmers have learned for the most part to do without fences. Rotations are so worked out that the pasture lot is not considered. Permanent pastures are established and the cattle confined to them. It is doubtful if the American farmer will soon get along without fences.

Photo by A. S. Rhodes.

Plant for creosoting fenceposts made of two discarded gasolene steel barrels.

Every woodlot contains trees large enough and strong enough that could be used for fence posts if it were not for the fact that the wood rots easily and quickly when put in the ground. Soft and hard maple, box-elder, ash, elm, beech, poplar, sycamore and other trees would be as serviceable as the best of cedar or locust if the wood were durable. Many of these trees, too, are fast growers. Cottonwood will grow to fence post size in 15 years or less, ash in 20 years, maple in 20 years, red oak in 25 years.

How can the trees that commonly grow in woodlots, but which do not produce durable woods, be utilized for fence posts? Can wood be made durable?

Wood decays because of the destruction of the cells by low forms of plant life called fungi, as has already been explained. When the spore of the fungus plant falls on moist wood it will start to grow. It sends out filaments into the wood consuming the fibers and causing decay. Like other forms of plant life, fungi require for their development heat, air, food and moisture. The reason a fence post decays most quickly at the surface of the ground is because there the conditions are most favorable for the growth of the fungi. Below the surface a post rots very slowly because of insufficient air, and above the surface the post rots slowly because the post dries out quickly after a rain. Any wood in contact with the ground or placed where it will be kept constantly moist will in time decay. If the fungus is deprived of any one of the factors of growth mentioned it cannot live and the wood will not decay. If wood is kept dry it will not decay. A table made even of a very perishable wood like birch or hickory will last indefinitely if kept in the house. Wood constantly covered with water cannot decay because of the lack of air for the development of the fungus plant. Bog oak logs are dug from the swamps in Ireland, where they have lain for centuries, and made into furniture. Wood decays very slowly in cold regions, because fungi need warmth. Wood that has been poisoned, so that the fungus is deprived of its food supply, will never decay as long as the poison remains in the wood. The methods of preserving wood are based on this principle of depriving the fungus plant of its food by poisoning the wood. This is the only factor that can be controlled practically in fence posts.

How the Life of Fence Posts can be Lengthened.—The drier a post is kept the longer will be its life. Draining the surface water away from a post will help to prolong its life. Piling stones, ashes or gravel around a post will help to drain away moisture and will keep down the weeds and grass that spring up about a post keeping the base moist and so subject to decay. Sometimes posts are set in cement or concrete. Because of the shrinking and swelling of wood with moisture this is not an effective method. The cement is cracked by the swelling of the posts and in dry times the posts will shrink and will not completely fill the cavity in the cement allowing the spores of fungi to be washed down between the wood and the cement. Coating a post with cement has the same objection.

It is possible to prolong the life of fence posts by preventing the entrance of fungi. Charring the ends of posts will cover them with a layer of charcoal through which the fungus can force its way only with difficulty. It also hardens the wood and may produce changes in the composition. In charring wood the danger lies in the wood checking badly and splitting open. The posts must be well seasoned. The charred portion must extend 8 to 10 inches above the surface of the ground when the posts are set. Wood that is kept well painted does not easily decay, because the layer of paint keeps out the fungus. If wood is wet when painted, however, it makes the wood especially liable to dry rot should that fungus gain entrance. Tar, pitch and various kinds of oils and other substances can be used for coating fence posts with considerable success. Surface coatings are of but temporary value. They are easily worn or broken off subjecting the wood to decay.

The most effective method of preserving fence posts is without doubt by destroying the food value, so that the fungus cannot live in it. The substances most commonly used for this purpose are poisonous salts, like copper sulphate, corrosive sublimate and zinc chloride, and oils such as crude oil and creosote. The use of salts for preserving fence posts is of value only in regions of scanty rainfall. In a moist climate salts are easily leached out of the wood. Creosote, which is a product of the distillation of coal tar, is the most important wood preservative. It not only poisons the wood so that fungi cannot live in it, but, being an oil, it tends to exclude moisture. Most of the patented wood preservatives on the market have creosote as their base. It can be obtained through hardware dealers.

METHODS OF TREATING FENCE POSTS WITH CREOSOTE

There are three methods of treating fence posts with creosote, the brush method, the dipping method and the open tank method.

The Brush Method of Treating Fence Posts.—The brush method consists in painting the posts with creosote. The posts are thoroughly dried and seasoned so that the creosote, which is applied hot, will be absorbed by the wood. The creosote is painted on the posts with an ordinary paint brush, and at least two coats should be given each coat being allowed to sink into the wood before another is applied. While the brush method will prolong the life of posts somewhat it is not an effective method because the creosote does not sink deeply into the wood and is easily leached out, and it is impossible to get the creosote into all cracks and season checks and defects, through each of which the fungus can gain entrance to the wood. The life of all timbers used in contact with the ground or where subject to damp conditions can be lengthened many years by this method.

The Dipping Method.—In this method the creosote is placed in a tank or barrel and the ends of the posts are dipped in the creosote. This is a more effective method than the brush method, but the creosote will not penetrate the wood very far. Both ends of the posts are sometimes dipped and often the tops of posts that have been treated by the open tank method are treated by dipping.

The Open Tank Method.—In the open tank method the posts are placed in a tank with creosote and heated to near the boiling point for several hours and then allowed to cool down in cold creosote. When the posts are heated in hot creosote the air inside the posts expands and a large part of the moisture in the wood is changed to steam and driven off. There is an outward pressure set up in the posts. After the posts have been heated for some time in hot creosote, if they are immediately placed in cold creosote, or if left to cool down in the creosote, the heated air inside the posts will contract and the steam will change to water forming a partial vacuum inside the wood. By atmospheric pressure the creosote will be forced deeply into the wood to take the place of the air and moisture forced out of the wood in heating. This forms a shell of creosoted wood about the post, all cracks, defects and season checks being filled. In some kinds of woods the creosote will sink an inch or more into the wood.

Preparation of the Posts for Treatment.—The posts should be thoroughly seasoned or air dried before treatment, as the presence of much moisture in the wood tends to prevent the entrance of the creosote. The posts should be piled loosely in some place out of the sun, but where the air will circulate freely, for several months. If piled in the sun the posts are apt to season check badly. The better the posts are seasoned the easier will be the treatment and the deeper the penetration. All the bark should be removed, especially the papery inner bark, as this prevents the creosote from entering the wood. Usually only about 40 inches of the lower end of the post is treated. This will bring the creosoted portion of the post 6 to 8 inches above the ground when the posts are set. It is at the surface of the ground where the fastest decay takes place. Unless the wood is very perishable it is not usual to treat the tops of the posts, but it can easily be done by inverting the posts in the tank when one end has been treated. Merely dipping the tops in hot creosote or using the brush process will have a very favorable effect. When the tops are not to be treated it is well to bevel them off so that they will shed rain.

Where posts are to be given a preservative treatment it is possible to use a much smaller post than is generally used. A post 5 inches in diameter is large enough and strong enough for most line fences. In using small posts a considerable saving of creosote can be made. Round posts are much preferred for the process, because sapwood will take the preservative much more readily than heartwood.

Equipment Necessary.—For treating a few fence posts each year a very satisfactory tank can be made from a discarded gasolene drum or barrel of sheet steel. The top should be cut out and the tank set up on bricks or stones, so that a fire can be built beneath. A stovepipe connected with the firebox will aid in giving a draft and carry the smoke away. An outfit of this kind at the most should cost but a few dollars. Such a tank will hold from eight to ten fence posts, 4 or 5 inches in diameter. Creosote being inflammable, the tank should not be set up near buildings.

Carrying out the Treatment in a Single Tank.—The creosote, which at ordinary temperatures is sometimes not in a liquid form, should be heated in the tank to about the boiling point. A very high temperature will tend to weaken the wood and a large amount of the creosote will be evaporated. The tank should then be filled with fence posts. Sufficient creosote should be used to cover at least the lower 40 inches of the posts. After heating for several hours the fire should be drawn and the posts left in the tank until cool. The posts will then be treated.

Two runs of posts can easily be treated in a day on the farm without interfering with the regular work. The tank can be filled with posts in the morning and a fire built that will last about two or three hours before burning out. The posts after being treated will cool down during the day and by night will be treated. A new run of posts can be put in, heated as before for two or three hours and allowed to cool down over night. In cold weather the tank will cool down fast enough to start a new run at noon. In this way three runs of eight or ten posts each can be treated without much effort every day.

The Two-Tank Method.—Where a large number of posts are to be treated the treatment can be given more quickly and economically by using two tanks. The posts are heated as before for several hours in hot creosote in the heating tank and then immediately transferred to a second tank containing cold creosote to cool for about an equal period. The result will be the same as when only one tank is used.

It is possible for farmers in the same region to cooperate in the purchase of a fence post treating outfit, thus reducing the initial cost. In this way larger equipment can be purchased and the work handled in a more economical way.

Length of Treatment.—The length of time necessary to heat the posts in hot creosote and to cool them in cold creosote to secure satisfactory penetration of the wood will vary with different kinds of wood. Some woods absorb creosote readily and are easily treated, while other woods require a long treatment. In general posts should be heated for from four to five hours in hot creosote and cooled an equal length of time. The depth of penetration can be known by cutting into a treated post. The more creosote a post absorbs and the deeper the creosote penetrates into the wood the longer the post will last, or, in other words, the longer the creosote will remain in the wood before being leached out by the rain. The greater the absorption, however, the greater the cost. To a farmer using his own time the cost of the creosote is the only real expense connected with the treatment. Pitch pine posts of the same size will absorb from 1 to 8 pounds of creosote, depending on the length of treatment, making the cost of creosote alone from about 2 cents to 16 cents. The quantity of creosote absorbed can be controlled by varying the length of time the posts are heated in hot creosote and cooled in cold creosote. A penetration of a half inch is sufficient for most kinds of posts. What is desired is a shell of creosoted wood about the posts that will remain for years and prevent the entrance of fungi. The heaviest treatment will, of course, give the best results.

Willow, soft maple, cottonwood and box-elder, common woods that last but 3 or 4 years in the ground or less untreated, will last 20 to 25 years or longer when treated at a cost of about 10 cents for both top and butt treatment. They should be heated for about 5 hours in hot creosote and cooled from 5 to 10 hours, with the exception of cottonwood which absorbs the treatment readily, and should be given a short treatment. It should be heated not more than 2 hours. Ash and red oak will last about 6 to 7 years in the ground untreated, and when given a treatment by heating for 6 or 7 hours in hot creosote and cooling for 10 to 12 hours, they should last 25 years or more.

ADVANTAGES OF TREATING FENCE POSTS WITH CREOSOTE

1. It permits the use of material the farmer has in his own woodlot that is not used at present because of the perishable nature of the wood.

2. By a preservative treatment the life of posts can be lengthened many times. Posts that last but 3 or 4 years can be made to last 20 to 25 years, thus saving not only the cost of new posts but the cost of replacing.

3. Smaller posts can be used when treated with creosote. Large posts are usually used because they will stand longer before decaying.

4. The cost of a post from the woodlot creosoted is much less than the cost of a durable post.

APPENDIX

SUGGESTED LIST OF PRACTICUM EXERCISES THAT COULD BE GIVEN IN CONNECTION WITH A COURSE IN FARM FORESTRY.

I. Tree study and identification. As many exercises as possible should be devoted to the study and identification of trees. The following method of conducting such exercises has been found to be satisfactory.

If the exercises are to start in the fall of the year study the hardwoods first and the conifers later in the winter; the opposite method should be followed in the spring. Take the students out to the tree to be studied and have them take notes on the following points:

1. Name, common and scientific.

2. Location of tree, open or forest grown, kind and character of soil and situation.

3. General form and appearance.

4. Method of branching.

5. Trunk diameter at breast height, clear length of bole and total height of tree (approximate).

6. Crown-width and length of crown, general shape and appearance, character of shade cast.

7. Bark color and characteristic markings.

Have each student secure a small branch with characteristic leaves and buds, and either study sitting around the tree or take to the class room. All observations made should be recorded by each student in notebooks, under the following heads, to be followed with each tree studied:

1. Twig.
2. Bud.
3. Leaves.
4. Fruit or seed.

1. Starting with the twig the instructor asks for observations on the twig. Have each member of the class study the twig and tell whatever is observed. Bring out the following points: size, color, thickness, smoothness, shiny or dull, markings, lenticles, annual nodes, leaf scars, etc.

2. Next take up the bud in the same way and bring out the following points: size by measuring, shape—conical, round, one-sided, long, short, pointed, blunt, scales, hairs, resin, sticky, smooth, where placed on the twig, and arrangement, comparison of terminal and lateral buds.

3. Then the leaf is observed. The following points should be brought out: Size and variation in size, general shape; lobes—number and form, round, sharp, color above and below, thickness; veins—arrangement, prominence, how far they extend, hairs above and below; stem—length, shape, how fastened to leaf and twig. With conifers the number of years the leaves remain on the trees should be noted.

4. The fruit and seed should be obtained where possible and studied in the same way, especially the form, size, shape, etc. Old seeds can usually be found on the ground under the trees.

After two or three trees have been examined in this way the method will be clear to both students and instructor. There is no better training for the observation. A tree studied in this way will never be forgotten. Do not allow one or two students to make all the observations. Only three or four trees should be examined in a practicum period. A few trees well studied are better than many trees skimmed over.

After studying the specimens, the leaf, twig, buds and

fruit can be drawn in the notebook free hand or by tracing the outline and filling in the detail by free hand, or the leaves can be dried between newspapers or regular drying boards and mounted. The instructor will supplement the notes with general remarks about the use of the tree for ornamental or forestry purposes and the value and uses of the wood. The woods of trees can be studied in much the same manner.

II. Heights of trees.—By the two-staff method. By comparing the length of the shadow of the tree and the length of the shadow of a pole of known length.

Record diameter at breast height and height of each tree measured as well as the name of the species.

III. Contents of individual trees.—In board feet by the use of the 10-foot pole and crosspiece and rule of thumb, and compare results by applying the Doyle and Scribner rules.

In cubic feet by the formula $\frac{B \times H}{}$

IV. Estimating timber.—Lay off one-sixteenth of an acre either in a square plot 52 feet on a side setting stakes at the corners, or by circular plots with 29½ feet radius, marking the trees on the edge of the circle. Use larger plots if time allows.

In board feet.—Estimate all the trees on the plot above 8 inches in diameter. Use 10-foot pole and crosspiece to get diameter and length of the butt log, estimate the diameters and lengths of upper logs. Apply rule of thumb and also log tables. Multiply the result by 16 to get the estimate for an acre.

In cordwood.—Find diameter outside the bark at breast height and total height, of all trees on the plot above a certain diameter, about 6 inches. Apply formula $\frac{B \times H}{2}$ to each tree. Multiply the estimate of the plot by 16 to get stand per acre.

V. Find area of woodlot. It will be better to lay off a small tract on level, open ground and find the area.

VI. Lay off an acre or half-acre plot. Make a list of all trees on the plot and find the percentage of the different species by dividing the number of trees of a certain kind by the total number of trees on the tract.

VII. Make a list of trees found growing only in the open, intolerant trees, and a list of those found growing under the shade of other trees, tolerant trees.

VIII. Note for different species of trees the time when the buds begin to swell in the spring, the time when the leaves are fully developed, the time of blossoming, the time when the leaves fall, and the time when the seed is ripe.

IX. Lay off a sample plot in the woodlot and have students write a description of the plot, kinds and percentages of trees, size and height of trees, condition of the trees, whether injured by fungus diseases or insects, the undergrowth, the reproduction, forest floors and the humus.

X. Have students mark with chalk or otherwise the trees that should be removed to increase the growth of the trees left or to improve the character of the woodlot.

XI. Have students study the market conditions, the products that are in demand and the prices of each delivered. Information can be secured from portable sawmill men, from lumber dealers, and from those that utilize the products or purchase them.

XII. Estimate all the products in a woodlot, or those in a certain portion, and figure what profit can be made by harvesting and selling the products.

XIII. Study the reproduction of trees in the woodlot to find under what conditions of soil and light the seeds germinate and the seedlings develop the best.

XIV. Visit a portable sawmill and scale logs with different log rules before they are sawed, and measure the amount of lumber that is actually sawed out of the logs, and compare the results. Determine the loss due to defects in logs.

XV. Visit a woodlot that has been heavily grazed and

another that has been recently damaged by fire, and note the injury to the trees and the forest floor.

XVI. Make a list of trees found growing in bottom lands, on lower and upper slopes and on ridges.

Method of Finding the Area of a Woodlot by the Use of an Ordinary Compass

Farm woodlots are usually more or less rectangular in outline or at least have straight sides, and areas should not be difficult to find. There are many ways of finding areas of tracts of land. The one described is one of the simplest methods and can be carried out with the ordinary compass reading from 0 to 360 degrees, that can be purchased for a small sum. A tape for measuring the sides of the woodlot is also necessary, although a stout string or a wire carefully measured can be used. Although a simple method, the one described, if carried out carefully, will give very accurate results.

The compass is set up at one corner of the woodlot (marked *A* in the diagram) so that the direction of one side can be seen. It is not necessary that the compass should set exactly on the corner. It should be placed where the direction of the side of the woodlot can be seen. The compass is for giving the direction of the side of the woodlot and nothing more. The compass should be set level, with the hand or pointer when at rest lying exactly over the point marked north. It will be found to be more convenient to set the compass on the ground rather than on a post or other elevation.

A string is now stretched over the compass in the direction of the side of the woodlot to be measured (*AB*), so that it passes directly over the center of the compass or the pin holding the needle. With an assistant holding the end of the string in the direction of the line to be run, and the other end held a short distance the other side of the compass, the direction of the side of the woodlot can be judged accurately.

The compass is then read from the center out to the edge of the compass in the direction of the line to be measured and the number of degrees where the string crosses noted. The distance *AB* or the side of the woodlot should then be

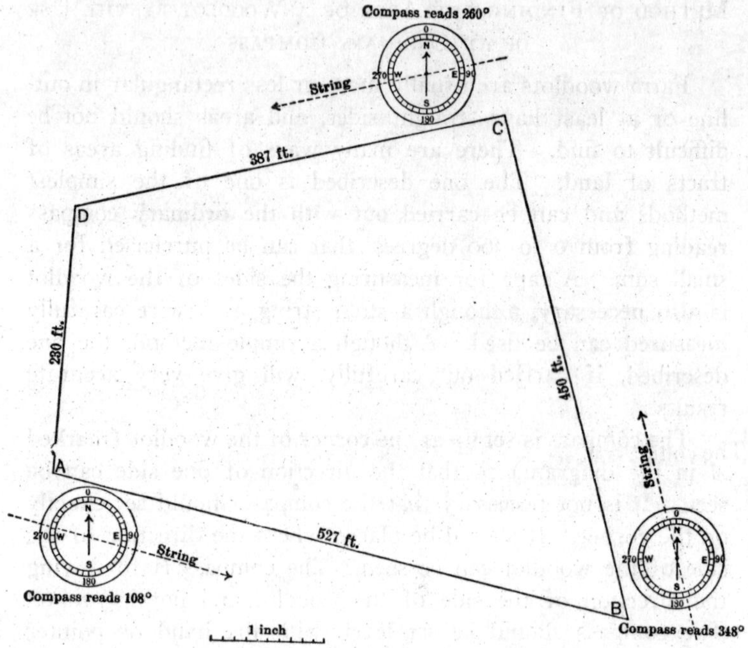

measured accurately in feet from corner to corner. The measurements can be recorded in a table as follows:

Line	Degrees	Distance
AB	108	527 feet
BC	348	450 feet
CD	260	387 feet
Etc.		

The compass is now set up at the corner *B* in the same manner as before, so that the needle on the compass will point exactly over the north point or zero point when at rest. The string is then stretched in the direction of the second side of the woodlot (BC) passing accurately over the center of the compass and the number of degrees noted. The side (BC) is then measured accurately in feet. The line BC, the number of degrees, and the distance are then recorded in the table. At each succeeding corner the same method should be followed. The direction of the line is found by means of the compass and string, and the distance measured accurately.

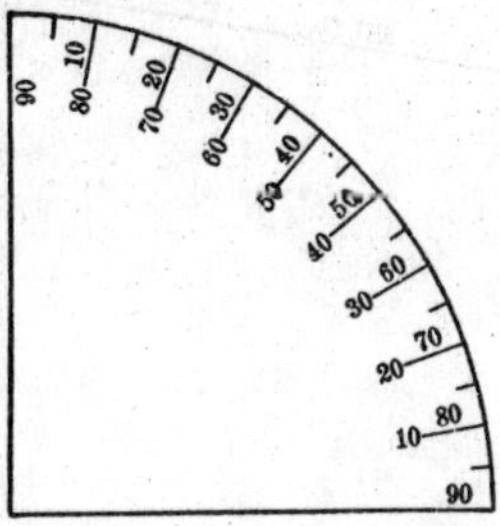

A map is now made of the woodlot from the data in the table. The method of drawing the map is similar in many respects to the measuring of the woodlot. The same processes are gone through with on paper. The same compass could be used, but it will be found more convenient to use a protractor such as is shown. A thin piece of paper can be laid over this protractor and traced and cut out for use.

In drawing the map the top of the paper is considered north and should be so marked. Then the bottom of the paper will be south, and the right hand edge east, and the left hand edge west. In drawing the map it is often a help to lay the paper on the table so that it lies in a north and south direction.

A point is now selected on the paper and marked A, to correspond with the first corner of the woodlot. A north and south line and an east and west line are now drawn through the point lightly as shown in the diagram. These correspond with the same points on the compass.

Using the data given in the table the direction of the line *AB*, or the first corner of the woodlot to the second corner, is 108 degrees. There are 360 degrees around the circle of a

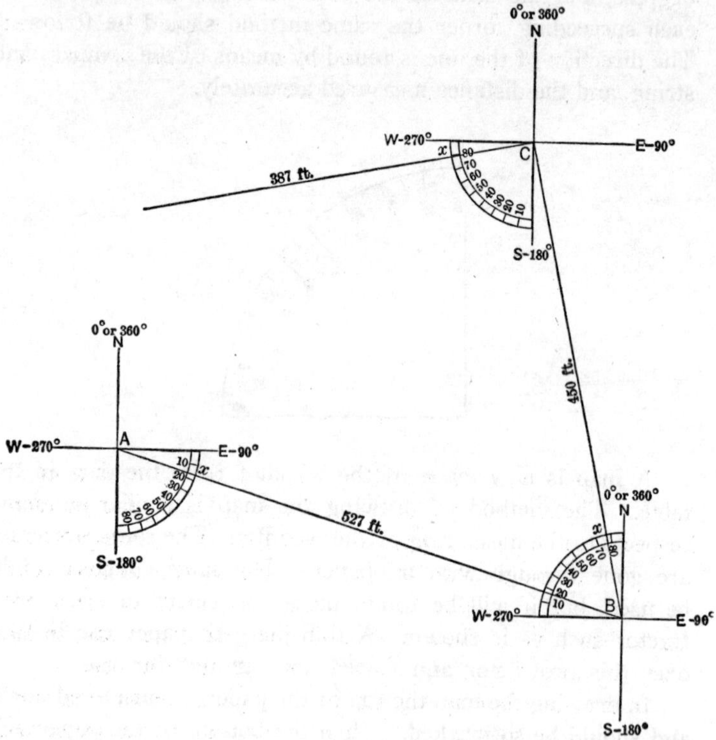

compass. From north to east as drawn on the paper through the point *A* there are 90 degrees. This will leave 18 degrees over to make the 108 degrees. Place the protractor so that it will lie between the east and south lines and count off 18 degrees and mark the point *x*. The line *AB* or the first side of the woodlot will pass from the point *A* through this point *x*.

From the table the length of the side *AB* is 527 feet. Using a ruler divided off into inches and tenths of inches and considering each inch on the ruler to be 100 feet, lay off on the line *Ax* extended, 5.27 inches. Mark this point *B*. The line *AB* will now represent the first side of the woodlot.

Through the point *B* as before draw a north and south line and an east and west line and proceed to map the side *BC* of the woodlot. From the table the direction as found by the compass is 348 degrees and the distance 450 feet. Proceeding as before from north to east is 90 degrees, from north to south is 180 degrees and from north around to west is 270 degrees. To find 348 degrees it will be necessary to pass 78 degrees beyond the west toward the north. Place the protractor so that it will lie between the north and west points as shown in the diagram and count off 78 degrees from the west point and mark the point on the paper *x*. The line *BC* will pass through this point *x* from the point *B*. Now with the ruler as before, allowing 100 feet to the inch, measure off 4.5 inches for the 450 feet and mark the point *C*.

Through the point *C* draw the east and west lines and north and south lines and lay off 260 degrees as noted in the table for the next line. This will necessitate passing 80 degrees beyond the south point. Lay the protractor between the south and west points and count off 80 degrees. The line *BC* will pass through the point found. Measure off 3.87 inches for the 387 feet and mark the point *D*. In like manner all the corners and sides of the woodlot are plotted on the paper.

When the last line is plotted, if the end of the line as drawn falls very close to the point *A* or the starting point, the work was well done. If the points do not come closely together some mistake was made either in the field work or in drawing the map, and the work should be done over again. Otherwise the last line should be drawn from the last corner to *A*, so that the map will close.

After the map is drawn and closed the area must be found. Using one side of the woodlot or map as a base, as shown in the diagram, mark it off in inches, carrying the line out if necessary beyond the map. At each inch mark erect lines perpendicular to the base line. Then divide these lines off into inches from the base line and draw lines through them parallel to the base line, thus dividing the entire map into square inches.

Each square inch represents an area 100 feet by 100 feet

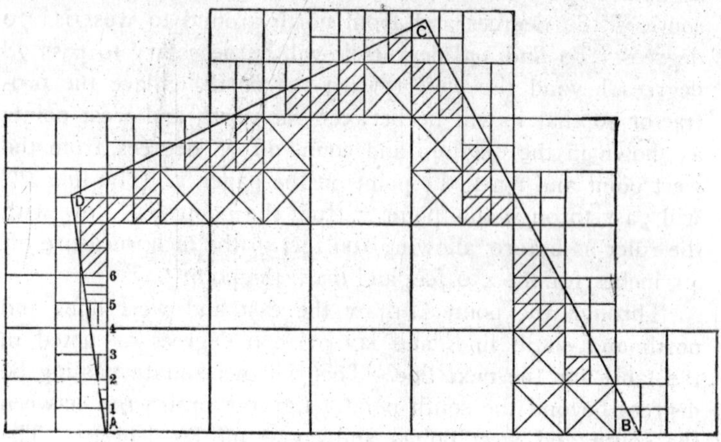

and contains 10,000 square feet. In the diagram there are 9 complete square inches, making 90,000 square feet. The areas of the imperfect squares must now be found. Divide each of these squares into quarters as shown, each of which will be 50 feet by 50 feet, containing 2500 square feet. Cross off every perfect quarter inch square. There are 10 of these in the diagram, making 25,000 square feet. Now divide each imperfect quarter inch square into quarters. These will be 25 feet by 25 feet each containing 625 square feet. In the diagram there are 25 of these perfect, making 16,875 square feet. The imperfect squares remaining around the edges of the map can now be combined to make perfect squares. For example, number 2 can be combined with number 6 to make a perfect square, and number 3 with number 5. Proceed in this way until the entire area of the map has been accounted for. The sum of all the different squares in the map will give the area of the woodlot in square feet. If this sum is divided by 43,560, the number of square feet in an acre, the number of acres in the woodlot will be found.

Bulletins and Circulars giving information on Farm Forestry.—Publications of the State Forestry Department and of the State Agricultural College on the establishment of farm woodlots and their care and protection, the kinds of trees to plant, the wood-using industries of the state, and the marketing of woodlot products.

Bulletins and Circulars of the United States Forest Service, a list of which can be obtained from the U. S. Department of Agriculture, Washington, D. C. The following relate to Farm Forestry and can be obtained from the Superintendent of Public Documents, Washington, D. C., by sending a postal money order for the amount:

U. S. Forest Service Bulletins:

36 Woodman's Handbook $0.25
42 Woodlot Handbook for owners of woodlands in southern New England .... .15
76 How to grow and plant conifers in the Northeastern States .... .10
78 Wood Preservation in the United States.......... .10
86 Windbreaks, their influence and value............ .30

U. S. Forest Service Circulars:

61 How to transplant forest trees................... $0.05
69 Fence post trees .... .05
96 Arbor Day, suggestions for its observance........ .05
99 Suggestions for forest planting on the semi-arid plains .... .05

117 Preservative treatment of fence posts............ .05
145 Forest planting on the western Prairies........... .05
195 Forest planting in the Northeast and Lake States.. .05

U. S. Department of Agriculture:

Bulletin 153—Forest planting in the Eastern United States $0.10

Year Book, Department of Agriculture:

Separate No. 651—The farm woodlot problem........ $0.10

Farmers' Bulletin:

134 Tree planting on rural school grounds........... $0.05
173 Primer of Forestry, Part I .... .05
358 Primer of Forestry, Part II .... .05
228 Forest Planting and Farm Management........... .05
423 Forest nurseries for schools.................... .05
582 Uses for chestnut timber killed by bark disease.... .05
622 Basket willow culture.......................... .05
715 Measuring and marketing woodlot products...... .05

TABLES FOR USE IN FINDING THE CONTENTS OF LOGS AND TREES

TABLE 1.—DOYLE LOG RULE

Length in Feet.	Diameter in Inches.																						
	6	7	8	9	10	11	12	13	14	15	16	17	18	19	20	21	22	23	24	25	26	27	28
	Board Feet.																						
8	2.0	4.5	8	12	18	24	32	40	50	60	72	84	98	112	128	144	162	180	200	220	242	264	288
10	2.5	5.6	10	16	23	31	40	50	62	75	90	106	122	141	160	181	202	226	250	276	302	330	360
12	3.0	6.8	12	19	27	37	48	61	75	91	108	127	147	169	192	217	243	271	300	331	363	397	432
14	3.5	7.9	14	22	32	43	56	71	88	106	126	148	171	197	224	253	283	313	350	386	423	463	504
16	4.0	9.0	16	25	36	49	64	81	100	121	144	169	196	225	256	289	324	359	400	441	484	530	576
18	4.5	10.1	18	28	41	55	72	91	112	136	162	190	220	253	288	325	364	406	450	496	544	596	648
20	5.0	11.3	20	31	46	61	80	101	125	151	180	211	245	280	320	361	404	452	500	551	605	661	720
22	5.5	12.4	22	34	50	67	88	111	137	166	198	232	269	309	352	397	445	496	550	606	665	726	792

TABLE 2.—SCRIBNER LOG RULE

Length in Feet.	Diameter in Inches.																						
	6	7	8	9	10	11	12	13	14	15	16	17	18	19	20	21	22	23	24	25	26	27	28
8	8	12	16	20	27	33	39	48	57	71	79	92	106	120	140	152	167	188	202	229	250	271	291
10	10	15	20	25	34	42	49	61	72	89	98	116	134	150	175	190	209	236	252	287	312	342	363
12	12	18	24	30	40	50	59	73	86	107	119	139	160	180	210	228	251	283	303	344	375	411	436
14	14	24	28	35	45	55	69	85	100	125	139	162	187	210	245	266	292	330	353	401	439	479	509
16	18	28	32	40	50	65	79	97	114	142	159	185	213	240	280	304	334	377	404	459	500	548	582
18	22	32	40	45	55	70	88	109	129	160	178	208	240	270	315	342	376	424	454	516	562	616	654
20	24	34	44	50	65	80	98	122	143	178	198	232	267	300	350	380	418	470	505	573	625	684	728
22	28	38	48	55	70	90	108	134	157	196	218	255	293	330	385	418	460	518	555	631	688	753	800

TABLE 3.—AREA OF CIRCLES IN SQUARE FEET

Diam. Ins.	Area Sq.ft.	Diam. Ins.	Area Sq.ft.	Diam. Ins.	Area Sq.ft.	Diam. Ins.	Area Sq.ft.	Diam. Ins.	Area Sq.ft.
1.0	.005	13.0	0.92	25.0	3.41	37.0	7.47	49.0	13.10
1.5	.012	13.5	0.99	25.5	3.55	37.5	7.67	49.5	13.37
2.0	.022	14.0	1.07	26.0	3.69	38.0	7.88	50.0	13.64
2.5	.034	14.5	1.15	26.5	3.83	38.5	8.08	50.5	13.91
3.0	.049	15.0	1.23	27.0	3.98	39.0	8.30	51.0	14.19
3.5	.067	15.5	1.31	27.5	4.12	39.5	8.51	51.5	14.47
4.0	.087	16.0	1.40	28.0	4.28	40.0	8.73	52.0	14.75
4.5	.111	16.5	1.48	28.5	4.43	40.5	8.95	52.5	15.03
5.0	.136	17.0	1.58	29.0	4.59	41.0	9.17	53.0	15.32
5.5	.165	17.5	1.67	29.5	4.75	41.5	9.39	53.5	15.59
6.0	.196	18.0	1.77	30.0	4.91	42.0	9.62	54.0	15.90
6.5	.230	18.5	1.87	30.5	5.07	42.5	9.85	54.5	16.20
7.0	.267	19.0	1.97	31.0	5.24	43.0	10.08	55.0	16.50
7.5	.307	19.5	2.07	31.5	5.41	43.5	10.32	55.5	16.80
8.0	.349	20.0	2.18	32.0	5.59	44.0	10.56	56.0	17.10
8.5	.394	20.5	2.29	32.5	5.76	44.5	10.80	56.5	17.41
9.0	.442	21.0	2.41	33.0	5.94	45.0	11.04	57.0	17.72
9.5	.492	21.5	2.52	33.5	6.12	45.5	11.29	57.5	18.03
10.0	.545	22.0	2.64	34.0	6.30	46.0	11.54	58.0	18.35
10.5	.601	22.5	2.76	34.5	6.49	46.5	11.79	58.5	18.67
11.0	.660	23.0	2.89	35.0	6.68	47.0	12.05	59.0	18.99
11.5	.721	23.5	3.01	35.5	6.87	47.5	12.26	59.5	19.31
12.0	.785	24.0	3.14	36.0	7.07	48.0	12.57	60.0	19.63
12.5	.852	24.5	3.27	36.5	7.27	48.5	12.83	60.5	19.96

TABLE 4.—WHITE PINE

Southern New Hampshire

From the Woodman's Handbook, U. S. Forest Service

Diameter, Breast-high.	Height of Tree (Feet).									
	30	40	50	60	70	80	90	100	110	120
	Volume (Board Feet)									
Inches										
5	8	12	15							
6	13	20	23	27	29					
7	18	28	34	39	44					
8	24	36	45	53	62					
9	32	44	56	69	81	93				
10	41	53	70	85	102	119	138			
11	....	63	84	103	126	147	168			
12	....	73	100	125	151	177	200	228	245	
13	....	84	117	148	180	210	238	270	293	
14	....	95	137	173	210	243	277	312	348	
15	....	105	158	200	241	282	321	362	406	
16	....	....	181	230	277	323	370	415	470	
17	....	....	209	261	313	368	421	471	540	
18	....	....	238	297	352	411	475	531	610	688
19	....	....	270	336	393	460	530	598	682	763
20	....	....	302	379	436	506	583	660	750	840
21	....	....	....	425	480	553	634	720	820	918
22	....	....	....	....	522	597	681	779	887	990
23	....	....	....	....	566	639	727	834	958	1065
24	....	....	....	....	....	674	769	889	1030	1135
25	....	....	....	....	....	706	809	942	1105	
26	....	....	....	....	....	737	846	994	1180	
27	....	....	....	....	....	....	....	1046		

TABLE 5.—RED OR NORWAY PINE

From The Woodman's Handbook, U. S. Forest Service

Diameter, Breast-high.	Height of tree (Feet).				
	60	70	80	90	100
	Volume (Board Feet).				
Inches.					
7	17	24			
8	29	38	50		
9	44	53	68	81	94
10	61	72	88	104	119
11	80	92	110	130	148
12	100	114	136	159	180
13	120	138	160	189	214
14	140	164	189	222	250
15	...	190	220	257	292
16	...	...	252	296	340
17	...	...	...	334	394
18	...	...	...	372	450

TABLE 6.—SPRUCE

Volume Table in Board Feet

From The Woodman's Handbook, U. S. Forest Service

Diameter, Breast-high.	Height in Feet.				
	40	50	60	70	80
	Volume (Board Feet).				
Inches.					
7	18	25	30	35	
8	29	38	45	53	
9	42	53	61	71	
10	58	67	78	91	
11	76	84	94	110	
12	96	100	112	130	
13	....	113	130	151	
14	....	129	148	172	194
15	....	....	166	195	219
16	....	....	186	219	245
17	....	....	208	244	275
18	....	....	....	272	305
19	....	....	....	308	343
20	....	....	....	346	400

TABLE 7.—SPRUCE

VOLUME OF UNPEELED PULP WOOD IN CUBIC FEET, SOUTHERN NEW HAMPSHIRE

From The Woodman's Handbook, U. S. Forest Service

Diameter, Breast-high.	Height of Tree (Feet)					
	40	50	60	70	80	90
	Volume (Cubic Feet)					
Inches.						
5	1.9	2.5	3.0			
6	3.5	4.2	5.2	6.4		
7	5.0	6.2	7.5	9.0		
8	6.6	8.4	10.0	11.7		
9	8.5	10.8	12.7	14.8		
10		13.5	15.6	18.0		
11		16.5	18.8	21.5		
12		19.5	22.3	25.4		
13			26.0	29.5	34.5	
14			30.0	34.0	39.5	
15			34.5	38.5	44.0	
16			39.0	43.5	49.0	
17			43.5	49.0	55.0	63.5
18			48.0	54.5	61.0	70.0
19			53.0	60.5	67.5	77.0
20			58.0	67.0	74.5	83.5
21				74.0	82.0	90.5
22				81.5	89.0	98.0
23				88.5	96.5	106.0
24				95.5	104.5	114.0
25				102.0	112.0	123.0
26				109.0	120.0	131.5
27					128.0	140.0
28					135.5	148.5

TABLE 8.—HEMLOCK IN BOARD FEET

From Report N. H. Forest Commission for 1906–7

Diameter, Breast-high.	Total Height of Tree (Feet)				
	30	40	50	60	70
Inches.					
6	5				
7	10	20	30	42	
8	17	28	39	50	
9	26	36	49	60	
10	36	46	59	71	86
11	47	58	72	86	103
12	60	72	86	103	123
13	....	88	104	124	148
14	....	107	125	147	173
15	....	126	148	172	204
16	....	148	171	200	240
17	....	....	197	233	281

TABLE 9.—BEECH IN PENNSYLVANIA, VOLUMES IN CUBIC FEET

From Bulletin 285, U. S. Dept. of Agriculture

Diameter, Breast-high.	Total Height of Tree (Feet)				
	70	80	90	100	110
	Volume of Logs Including Bark (Cubic Feet)				
Inches.					
8	8.8	10.1	11.3		
9	11.4	13.0	14.6		
10	14.3	16.4	18.4	20	
11	17.7	20.0	23.0	25	
12	21.0	24.0	28.0	31	34
13	26.0	29.0	33.0	37	40
14	30.0	34.0	39.0	43	47
15	35.0	40.0	45.0	50	55
16	......	46.0	52.0	57	63
17	......	52.0	58.0	65	71
18	......	58.0	65.0	72	80
19	......	64.0	72.0	80	88
20	......	70.0	79.0	88	96
21	......	......	86.0	95	105
22	......	......	92.0	103	113
23	......	......	99.0	110	122
24	......	......	106.0	118	130
25	......	......	113.0	125	138
26	......	......	......	133	146
27	......	......	......	141	155
28	......	......	......	149	164
29	......	......	......	157	173
30	......	......	......	164	181

TABLE 10.—VOLUME OF CHESTNUT IN CUBIC FEET

From Bulletin 96, U. S. Forest Service

Diameter, Breast-high.	Height of Tree (Feet)							
	20	30	40	50	60	70	80	90
	Volume							
Inches.	Cu.ft.	Cu.ft.	Cu.ft.	Cu.ft.	Cu.ft.	Cu.ft.	Cu.ft.	Cu.ft.
10	....	....	10.5	12.5	14.5	16.6	17.8	19.0
11	....	....	12.8	15.3	17.9	20.7	22.5	24.3
12	....	....	15.4	18.4	21.5	25.0	27.4	30.0
13	....	....	18.2	21.7	25.2	29.1	32.1	35.3
14	....	....	....	25.0	29.2	33.8	37.7	41.3
15	....	....	....	28.8	33.6	38.8	43.4	48.0
16	....	....	....	32.6	38.1	44.0	49.5	55.0
17	....	....	....	36.5	42.7	49.5	56.0	63.0
18	....	....	....	40.5	47.4	55.5	63.0	70.5
19	....	....	....	44.3	51.4	61.5	70.0	79.0
20	....	....	....	49.0	58.5	68.5	78.0	88.0
21	....	....	....	....	65.0	75.5	86.0	97.0
22	....	....	....	....	71.0	83.0	95.0	106.5
23	....	....	....	....	78.0	91.0	104.0	117.0
24	....	....	....	....	85.5	99.5	114.0	128.0
25	....	....	....	....	93.0	108.5	124.0	140.0

TABLE 11.—VOLUME OF CHESTNUT IN LUMBER

From Bulletin 96, U. S. Forest Service

Diameter, Breast-high.	Height of Tree (Feet)				
	50	60	70	80	90
	Volume in Board Feet				
Inches.	Bd.ft.	Bd.ft.	Bd.ft.	Bd.ft.	Bd.ft.
9	10	15	22		
10	26	32	40	48	56
11	42	50	58	69	80
12	58	68	78	92	107
13	74	87	100	116	133
14	92	107	122	141	160
15	110	127	147	168	190
16	129	150	172	196	222
17	......	174	200	226	255
18	......	198	227	257	291
19	......	223	257	292	328
20	......	250	288	327	368
21	......	276	318	363	409
22	......	305	350	400	451
23	......	333	385	440	493
24	......	363	420	479	538
25	......	396	457	520	583

TABLE 12.—VOLUME OF RED, BLACK, AND SCARLET OAK, IN CUBIC FEET

Taken from U. S. Forest Service Bulletin 96

Diameter, Breast-high.	Height of Tree (Feet)						
	20	30	40	50	60	70	80
	Volume						
Inches.	Cu.ft.	Cu.ft.	Cu.ft.	Cu.ft.	Cu.ft.	Cu.ft.	Cu.ft.
2	0.3	0.4					
3		0.9	1.0				
4		1.5	2.0	2.6			
5		2.4	2.9	3.6			
6			4.2	5.0	6.0		
7			5.7	6.6	7.6	8.9	
8			7.0	8.5	10.2	12.0	
9			8.7	10.9	13.0	15.2	17.3
10			10.5	13.1	15.7	18.4	21.0
11			13.0	16.0	19.0	22.1	25.2
12			16.4	19.7	22.8	26.2	29.3
13			21.0	24.3	27.7	31.0	34.3
14			25.3	29.0	32.6	36.3	40.0
15			30.2	34.2	38.1	42.0	45.8
16				40.0	44.1	48.2	52.4
17						54.6	59.1
18						61.0	66.0
19						68.0	74.0

TABLE 13.—VOLUME OF RED, BLACK, AND SCARLET OAKS IN LUMBER

Taken from U. S. Forest Service Bulletin 96

Diameter, Breast-high	Height of Tree (Feet)			
	50	60	70	80
	Volume			
Inches.	Bd.ft.	Bd.ft.	Bd.ft.	Bd.ft.
9	14	17	21	
10	24	30	38	49
11	35	45	56	68
12	48	60	75	89
13	62	77	94	113
14	78	96	117	140
15	95	117	141	168
16	114	140	168	198
17			195	229
18			224	263
19			254	300

TABLE 14.—VOLUME OF WHITE AND CHESTNUT OAK, IN CUBIC FEET

From Bulletin 96, U. S. Forest Service

Diameter, Breast-high.	Height of Tree (Feet)							Total Basis.
	20	30	40	50	60	70	80	
	Volume							
Inches.	Cu.ft.	Cu.ft.	Cu.ft.	Cu.ft.	Cu.ft.	Cu.ft.	Cu.ft.	Trees.
2	0.3	0.4						26
3	.....	0.9	1.0					55
4	.....	1.5	2.0	2.6				54
5	.....	2.4	2.9	3.6				46
6	.....	.....	4.2	5.0	6.0			38
7	.....	.....	5.7	6.6	7.6	8.9		36
8	.....	.....	7.4	8.7	10.3	12.2		10
9	.....	.....	9.3	11.1	13.5	16.0	18.9	2
10	.....	.....	11.2	13.7	16.5	19.5	22.6	3
11	.....	.....	13.3	16.5	19.8	23.2	26.8	14
12	.....	.....	15.7	19.3	23.1	27.2	31.4	1
13	.....	.....	.....	22.4	26.9	31.5	36.3	2
14	.....	.....	.....	25.9	31.0	36.3	41.6	1
15	.....	.....	.....	.....	35.5	41.7	48.1	
16	.....	.....	.....	.....	40.5	47.5	55.0	
Total	.....	.....	.....	.....	.....	.....	.....	293

Volumes include stem and topwood, with bark, up to a minimum diameter of 2 inches. Average stump heights vary from 6 inches for small trees to 21 inches for large ones.

TABLE 15.—WHITE OAK

VOLUME IN BOARD FEET BY THE DOYLE-SCRIBNER RULE, SOUTHERN APPALACHIAN REGION

From the Woodman's Handbook

Diameter, Breast-high.	Height of Tree (Feet)						
	60	70	80	90	100	110	120
	Volume (Board Feet)						
Inches.							
14	55	70	90	115			
15	80	95	115	140			
16	100	115	140	165	195		
17	125	140	165	195	225		
18	150	165	190	225	260		
19	175	190	220	255	295		
20	200	215	250	285	335	400	
21	225	245	280	320	375	450	
22	250	275	310	355	415	495	
23	280	305	340	390	455	545	
24	305	335	375	425	500	595	
25	335	365	410	465	545	645	
26	....	400	445	505	590	695	865
27	....	435	480	545	635	750	920
28	....	470	520	585	685	805	980

TABLE 16.—SUGAR MAPLE IN PENNSYLVANIA, VOLUMES IN CUBIC FEET

From Bulletin 285, U. S. Department of Agriculture

Diameter, Breast-high.	Total Height of Tree (Feet)				
	70	80	90	100	110
	Volume of Logs Including Bark (Cubic Feet)				
Inches.					
10	13.6	15.6	17.5	19.5	
11	16.7	19.1	21.0	24.0	
12	20.0	23.0	26.0	29.0	32
13	24.0	27.0	31.0	34.0	37
14	28.0	32.0	36.0	40.0	44
15	32.0	37.0	42.0	46.0	51
16	....	43.0	48.0	53.0	59
17	....	49.0	55.0	61.0	67
18	....	55.0	62.0	69.0	76
19	....	62.0	70.0	78.0	85
20	....	69.0	78.0	87.0	95
21	....	....	89.0	96.0	106
22	....	....	96.0	106.0	117
23	....	....	104.0	116.0	128
24	....	....	113.0	126.0	139
25	....	....	122.0	136.0	149
26	....	....	....	145.0	160
27	....	....	....	155.0	171
28	....	....	....	164.0	181

TABLE 17.—SECOND GROWTH SUGAR MAPLE, VOLUMES IN BOARD FEET

From Bulletin 176, Vermont Agri. Exp. Station

Diameter, Breast-high.	50	60	70	80
Inches.	Bd.ft.	Bd.ft.	Bd.ft.	Bd.ft.
7	18	18	18	36
8	24	36	42	42
9	42	50	62	74
10	....	68	74	82
11	....	90	90	96
12	....	100	126	126
13	....	118	144	160
14	....	....	180	186

TABLE 18.—SECOND GROWTH YELLOW POPLAR

VOLUME IN BOARD FEET BY THE SCRIBNER RULE, FAIRFAX COUNTY, VA.

From Woodman's Handbook, U. S. Forest Service.

Diameter, Breast-high.	Height of Tree (Feet)						
	40	50	60	70	80	90	100
	Volume (Board Feet)						
Inches.							
7	2	7	12	16			
8	5	11	17	23	29		
9	9	17	25	32	41	48	
10	15	25	35	45	56	67	78
11	22	36	48	61	74	88	100
12	32	50	65	80	94	110	123
13	......	66	84	101	117	134	148
14	......	......	106	124	142	160	177
15	......	......	129	150	172	191	212
16	......	......	......	179	202	225	250
17	......	......	......	210	236	264	288
18	......	......	......	......	274	304	328
19	......	......	......	......	318	346	374
20	......	......	......	......	......	395	428